W. T. Greene

Parrots in Captivity

W. T. Greene

Parrots in Captivity

ISBN/EAN: 9783337151256

Printed in Europe, USA, Canada, Australia, Japan

Cover: Foto ©berggeist007 / pixelio.de

More available books at **www.hansebooks.com**

PARROTS IN CAPTIVITY.

BY

W. T. GREENE, M.A., M.D., F.Z.S., Etc.,

Author of "The Amateur's Aviary of Foreign Birds," &c.

WITH NOTES ON SEVERAL SPECIES BY THE HON. AND REV. F. G. DUTTON.

* *

ILLUSTRATED WITH COLOURED PLATES.

LONDON:
GEORGE BELL & SONS, YORK STREET, COVENT GARDEN.
MDCCCLXXXIV.

Contents.

	PAGE
Rose-Hill or Rosella Parrakeet	1
Pale-Headed or Mealy Rosella	6
Yellow-Rumped Parrakeet, or Broadtail	8
New Zealand Parrakeet	13
Golden-Crowned Parrakeet of New Zealand	16
Blue Bonnet Parrakeet	21
Blood- or Red-Rumped Parrakeet	23
Many-Coloured Parrakeet	26
Beautiful or Paradise Parrakeet	29
Swift Parrakeet, or Lorikeet	33
Passerine or Blue-Winged Parrakeet	37
Grey Parrot	41
Senegal Parrot	57
Hyacinthine Macaw	61
Military Macaw	65
Red and Blue Macaw	69
Blue and Yellow Macaw	75
Illiger's Macaw	81
Carolina Parrot, or Conure	84
Golden-Crowned Conure, or Half-Moon Parrakeet	88
White-Eared Conure	92

CONTENTS.

	PAGE
Festive Amazon Parrot	94
Dufresne's Amazon	96
Blue-Fronted Amazon	98
Double-Fronted or Le Vaillant's Amazon	104
Red-Vented Parrot	108
Dusky or Violet Parrot	110

INTRODUCTION.

THE height of our ambition in the bird-keeping line for many years was to have a Parrot of our own: we were not at all particular as to the species, or as to the colour of the bird, whether grey, green, or white, whether long or short-tailed, we did not greatly care, so that the creature could be called a Parrot, and was able to talk.

A friend of ours possessed a fine Green Parrot that had taken, why we know not, an inveterate dislike to us, and invariably accosted us when we entered the room where it was caged with the uncomplimentary exclamation: "Get out, you dirty brute!" but in spite of this unmerited reproach, for at that time we were especially solicitous about our appearance, we would have given anything to own the bird, than which we have seldom, if ever, heard one that pronounced quite a number of sentences so distinctly.

Then another acquaintance had a very clever Grey Parrot, that almost spoke as plainly as our green detractor, if with a less extended vocabulary, and another had a white Cockatoo of Australian extraction, which, if it had not a great deal to say for itself, was an accomplished acrobat, and withal so handsome, as it erected its bright sulphur crest, that one could not but overlook its linguistic deficiencies in favour of its charming personal appearance.

Green Ring-necked Parrakeets, Australian Broadtails and Cockatiels, and the marvellously beautiful Rosella and Pennant, the tiny Love-birds, and the Dove-coloured Parrot with its breast of rose—how we envied their owners, or at least longed to have one like them of our own.

Yet, had our wish been gratified then, it is probable we should not have been able to retain the object of our desires very long, for we were absolutely ignorant of the treatment necessary for preserving one of these birds in health.

INTRODUCTION.

Bread and milk we had been assured was the only proper food for a Parrot, and upon that strange diet not a few of those with which we had been acquainted were habitually fed, and some of them even survived its administration for a considerable time, thanks, no doubt, to an exceptionally fine constitution.

"What is the proper time of year to buy a Parrot?" is a question that we are very frequently asked, and to which we reply, In the summer time; that is to say from June to the end of August, or September if the weather is exceptionally fine: if you buy one during the cold season it will be apt, even if you carry it home yourself, to take a chill upon being brought out of the dealer's stuffy shop, for, mark you, all bird-dealers' shops are stuffy; while if it is sent to you from a distance, it will, if alive when it reaches you, be certain to die in a few days, or, at the outside, weeks after you have received it, from bronchitis or inflammation of the lungs, the foundation of which has been laid on the journey. Let a bird be ever so well packed up when it leaves the hands of the dealer, it will be certain to be exposed to the cold through the curiosity of the railway porters, who seem perfectly unable to let a bird pass through their hands without personal inspection, and if to the exposure thus entailed upon it, we add a sojourn of some hours in a cold railway van, the wonder is not that so many of the birds sent by rail die, but that any of them survive. We have known instances where birds had been securely packed in boxes, the lids of which were screwed down, and the air-holes protected by perforated zinc and muslin, which on reaching their destination were found to have had the protecting gauze torn away, and even the screws taken out, so that the porters should see what sort of a thing they were handling. In summer-time this curiosity on the part of the railway officials is not likely to be productive of such serious consequences as in winter; but, nevertheless, we would advise our readers never to send to a distance for a bird if they can by any possibility buy one near at hand, and carry it home themselves. We would prefer to give £2 for a bird that we could inspect previous to purchase, and take away with us, rather than half that amount for one that would have to be sent to us by rail.

Should it, however, be decided to buy a Parrot in the winter, or spring, the purchaser must take it home in a snug box, and place the cage it is to inhabit in a warm room, taking the precaution to cover it over on top and three sides with a thick baize covering, to place it below the gas, and having ascertained by inspection what the dealer had been feeding it on, to give it the same food; after having had the bird for some time, say a week or so, during which period it must on no account be exposed to any cold, but the temperature be kept even, say about 70°, the baize cover may be gradually removed, and the bird accustomed to the ordinary temperature of the house, but in any case it will be well to cover it up at night.

We now come to a point which is of the utmost importance to observe, namely, the treatment of a newly-purchased Parrot during the night. Many people are apt to forget that when the light and the fire are out during the night in a room which had been very warm during the day-time, the temperature falls a considerable number of degrees, and the bird gets chilled, and too often dies. If the owner has a fire in her bed-room, let her take the cage up with her when she is going to bed herself, but if not, let the fire be well banked up before retiring to rest, and let no housemaid enter the room in the morning before the mistress of the bird: for housemaids, as a rule, even when not actually inimical to Parrots, are utterly regardless of them, and have a habit of throwing up the window, even on the coldest morning, and letting the chilly outside air blow in keenly on the bird, which feels it all the more from having been kept snug and warm before; and it is this *mauvais quart d'heure* in the morning before the fire is lit, and while the room is being "dusted", that is fatal to so many Parrots, whose owners wonder how "poor Polly" can possibly have taken cold.

It may here be asked, "Supposing my Parrot to have taken cold, what is the best course to adopt?" Well, in ninety-nine cases out of one hundred, the bird will, in spite of every care, gradually get worse and die; but in the hundreth "she" has an extraordinarily good constitution, and if kept in an equable temperature of from 70 to 75°, and fed on soft food, soaked bread and boiled maize, for instance, and has tepid or lukewarm water to drink, she will recover: but prevention is better than cure; see that your bird does not catch cold, and you will not have the trouble of nursing it, and the grief of seeing it suffer, and probably die.

Avoid purchasing a Parrot in the winter time, and especially avoid having one sent to you a long distance by rail; but curb your natural impatience to become the owner of a Parrot until summer, and, where practicable, carry home your prize from the place where you have bought it; and, having got it home, see that is exposed to no draught, even in summer, but is gradually accustomed to the ordinary variations of temperature in your house.

INTRODUCTION.

In the matter of food for Parrots, the public has yet very much to learn: the traditional bread and milk, like the customary hard-boiled egg for Canaries, is a mistake: Parrots are graminivorous, that is to say, vegetable feeders; a few of them, it is true, live almost exclusively on honey, and a less number are partially insectivorous in their habits, but no known bird of this genus touches flesh in its wild state, the rumoured partiality of the *Nestor notabilis* for live lambs notwithstanding. Therefore to feed a bird of this description on animal food, such as milk, butter, bits from the table, bones and so on, is to force it to partake of an unnatural diet, which is certain, sooner or later, to produce disease, and ultimately to destroy the bird. To argue that because a Parrot appears to enjoy such an abnormal course of feeding, it is good for it, is about as sensible as to say that children love sweetmeats, and may, without endangering their health, be fed largely, if not exclusively, upon them, for Parrots, like children, are not always, indeed never, able to discriminate between those things that are suitable for them, and those that are injurious.

The larger Parrots require large seeds, such as maize, oats, dari, buckwheat, and dry biscuits (without milk or butter), nuts of various sorts, Brazilian, cob, and especially monkey nuts; and the smaller varieties, canary seed, millet, hemp, and a few oats occasionally. As all these birds in their wild state subsist more or less on unripe, or at least soft seeds, and fruit of different kinds, apples, pears, grapes, and oranges may be included, with discretion, in their bill of fare; and a portion of the different kinds of seeds that are offered to them should always be supplied either boiled or soaked in water until soft.

Again, these birds should always have access to water: it is true that many of them will exist for a long time, even on a diet of dry seed, without drinking, but Parrots in their wild state always drink, if even some of them confine their potations to sucking the drops of dew off the leaves and grass; and in captivity water is even more necessary to maintain them in health, for the staple of their food is dry, and they have not the chance of sipping the pearly drops of dew. Deprivation of water produces indigestion, causes heat and irritation of the skin, and often leads to the poor bird stripping itself bare of feathers.

Although a Parrot has a strong beak, it has no teeth, and is unable to masticate its food, swallowing the smaller seeds whole, that is after having stripped them of their husks, and the larger in little fragments, which require softening in the crop, and triturating in the muscular stomach or

INTRODUCTION.

gizzard before they can be digested, and serve for the nourishment of the bird. To supply the want of teeth, Nature has taught the Parrot to swallow a certain number of small, sharp-edged stones, which effectually reduce the food to a pulp, and prepare it for absorption by the glands of the stomach and intestines: yet how few owners of Parrots ever think of supplying their pets with such small artificial teeth as we have alluded to.

Many Parrots, especially the Australian species, appear in their wild state to evince a preference for brackish, or slightly salt, water, over fresh; yet we suppose it has not occurred to one Parrot keeper out of a thousand to supply the bird he or she owns with a morsel of rock salt: in fact many people look upon salt as rank poison for any bird, a belief in which we were strictly educated, but which we now know to be without foundation in fact: salt, instead of being injurious to Parrots, is very beneficial, and should always be supplied to them: very sparingly, of course, at first, but when the Parrot has got over the novelty of the thing, like the pastrycook's errand-boy, it may be safely trusted with a lump in its cage, it will not take more than is good for it.

A Parrot is naturally an extremely active and lively bird, and should never be kept in a small cage, in which not only it is seen to disadvantage, but is apt to injure its plumage, especially its tail, and the ends of the flight feathers in the wings: even when kept in a large-sized cage, it should, when practicable, be allowed a fly, every now and then, about the room, which it will much enjoy, and when its brief period of liberty has expired, a *bonne bouche*, in the shape of a morsel of biscuit, a nut, or a piece of apple, will soon lure it back to its domicile, to which, after a while, it will return of its own accord, when it is tired of rambling about.

Road grit, well washed to free it from dust and other impurities, is the best thing with which to supply Parrots in lieu of a carpet for the floor of their abode, and the absurd grillage with which the cages of these birds are invariably provided when purchased new, should be immediately removed; for, instead of answering any useful purpose, it is positively injurious, hindering the Parrot from reaching the sharp grit on the bottom of the cage; and was, doubtless, invented by some lazy owner, who objected to the trouble of cleaning out "Polly"'s habitation as frequently as he should have done: which brings us to the subject of cleanliness.

This virtue, we have been assured on very high authority, is akin to Godliness, but without going quite as far as that, we are bound to say that it is of the utmost importance, if the health and beauty of the captive

Parrot are to be taken into consideration. When refuse food and the droppings of the creature itself are allowed to accumulate on the floor of a bird's cage, bad smells arise from fermentation, not to say putrefaction, and, it is scarcely necessary to insist on such a point, are apt to give rise to low fever, and other serious ailments: the perches, too, must be frequently scraped or washed, or the Parrot, whose legs are so short that the breast feathers rub on what the bird sits on, will soon present a bedraggled, poverty-stricken appearance, that will cause it to neglect the rest of its person, and finally to become an object of disgust to those who see it, instead of, as it otherwise would be, "a thing of beauty" and "a joy for ever."

Parrots are great dandies, and are fond of preening and arranging their beautiful dress, but let this once become seriously soiled, they will give up the care of their toilet in despair, and degenerate into hopeless slatterns and slovens. In order to enable them to maintain the natural beauty of their coat, these birds, in their wild state, bathe freely, and should always be provided with the means of "tubbing" when they please in captivity. Some of them will not wash themselves in a cage, but as soon as they are allowed to fly about a room, will pop into a pan of water, if one is placed where they can readily reach it, and give themselves a good washing, after which they will sit in the sun, and pass every feather through their beak, until they are as neat as a new pin, and as glossy as if they had just come out of the woods.

Some birds however, in all probability individuals who have been brought up by hand from the nest, cannot be prevailed upon to wash themselves at all; in which case it will be necessary to give them, not a good scrubbing, but a gentle shower-bath of luke-warm water from a garden syringe, every now and then: and after the first time or two they will cease to be afraid, and even like and look for their washing, which, we need scarcely observe, must only be done in warm weather. Or the cage may be placed out of doors during the continuance of a genial summer shower, which "Poll" will much appreciate, holding out his wings, and spreading out his tail to catch the falling drops; this we have seen our acclimatised Parrakeets doing, even in winter time, in the out-of-doors aviary where we keep them all the year round.

Never put a Parrot in a round cage: nothing mars the beauty of its plumage so much, for in turning about it is obliged to press against the bars of the cage in whichever direction it moves, and so the feathers get frayed and broken; a square cage with two perches in it, one placed crosswise above the other, is the proper abode for a Parrot, and the larger the dwelling, the better and more healthy will be the bird.

A cross-bar stand to which the creature is chained by the leg is perhaps preferable to a cage for the Macaws and larger Cockatoos, but care must be taken that the part upon which the bird sits is not cased with tin, but made of wood, the ends of which only should be covered with metal: but a perch of iron or zinc is too cold for the feet of a Parrot, who gets cramp, and pains in his limbs from sitting on such an unnatural kind of perch, which a considerate owner will no longer compel him to use, when he knows what suffering it entails upon the unfortunate bird.

Parrots, as a rule, have as much individuality, not to say character, as human beings, each has its peculiar idiosyncrasy, and no hard and fast rule can be laid down for their management, as each several bird must be studied and treated according to the disposition it displays: this is particularly true of the large Parrots, including the Cockatoos, but the smaller species, namely the Parrakeets and Love-birds, thrive better in an aviary than they do in a cage. These small creatures very seldom become as much attached to their owners as their larger brethren frequently do, and we have never known an instance in which they did not prefer the society of a member or members of their own race to that of the master or mistress who had bought and cared for them; whereas, the contrary rule very frequently obtains with regard to the large Parrots. In any case a bird that may be comparatively tame and gentle when kept in a cage, or chained to a stand, by itself, is very apt to become wild, even savage, when placed in the society of a companion of its own kind, although this is by no means invariably the case; and, as we said before, the idiosyncrasy of the bird must be considered in this respect.

Some Parrots and Parrakeets will become so tame that, especially in the country, they may be permitted to enjoy almost perfect liberty in the garden, returning regularly to their owner's call, or at all events when prompted by the demands of appetite, for which reason it is always well to let them out, at least at first, before they have had a meal, and to hold out the sweetmeat, in a manner of speaking, to them when it is wished that they should return to their cage.

Some species are gifted with more, much more, of the homing instinct than others; the Parrots proper and the Lories, for example, far exceeding the old world Parrakeets, such as the Ring-necked and Alexandrine, in this respect, while the Australian broad-tails come about midway between the other two families. The Australian Parrakeets will generally return to their mates, if the latter, whether male or female, be placed where they can see and hear each other, but the Long-tailed Indian Parrakeets, once they get out of their cages, simply fly straight away, like an arrow from a bow, rejoicing, no doubt, in their new-found freedom, and utterly oblivious, apparently, of the guid wife or the guid man at home.

The Love-birds, too, have no idea of returning to their prison, and, once they escape, are very seldom seen again by their owners or their mates, to whom they really bear nothing like the affection with the possession of which they are popularly credited.

The Grey Parrots, the large Green Parrots, commonly called Amazons, the Macaws, and the Cockatoos are the best homers, then come the beautiful Indian Lories, and the Australian Grass Parrakeets; all the remaining species of the race are not in this respect to be depended on; once their liberty is regained, death is by them preferred to a return to captivity, even when a former mate calls to and tries to entice them back again.

We have often been asked which is the best way to teach one of these birds to speak, and have replied that there is no royal road: patience and perseverance alone will succeed, though some of them it must be admitted are much more ready learners than others: few of the hen birds, for instance, ever become accomplished linguists, although to this rule, as to every other, there are certain exceptions; but as, in the case of small birds, the gift of song is chiefly confined to the male, so in the Parrot tribe the capacity for learning to repeat articulate sounds is not usually the prerogative of the gentler sex.

For a talker then, select a male, and repeat to him slowly and distinctly the word or sentence it is wished to teach him; the bird will probably take it up word by word, not always beginning at the beginning, but occasionally in the middle of the phrase, as in the case of a Jardine Parrot belonging to the Hon. and Rev. F. G. Dutton, which was learning "Polly put the kettle on", and began by repeating "kettle, kettle", and gradually added the other words until it had learned to say the sentence correctly.

Captive Parrots, poor things! are frequently the subjects of disease, too often induced by errors of management on the part of their owners, who, not knowing any better, pamper their pets with unsuitable dainties, and then wonder why their birds should be ill and die. Some of these complaints we have already touched upon, but there are others, such as consumption, which can only be cured if taken in hand at the very commencement of the attack, and are much easier to prevent than to cure: exposure to a low temperature, and insufficiently nutritious diet, are the exciting causes of this complaint, for which the remedies are continuous warmth and appropriate and nutritious food: many drugs and nostrums have been recommended, but we have not faith in any of them. The symptoms of consumption are gradual emaciation, distaste for food, shivering, listlessness, sometimes a little cough, and in the latter stages diarrhœa: when the last complication has set in the case is hopeless. Another complaint, often fatal with newly-imported birds, is fever, generally of a typhoid character, which is almost incurable: a bird so afflicted is inordinately thirsty, drinking as much, in some cases, as a pint of water per diem. In slight attacks we have found dilute, or aromatic sulphuric acid in the proportion of ten drops to the ounce of water productive of benefit; the diet should be nutritious,—sponge-cake, a little bread and milk, (which latter article is only admissible as a medicine, or for very young subjects) and, where there is a tendency to dysentery, that is to say blood-stained evacuations, mutton broth in which rice has been cooked.

Pneumonia, or inflammation of the lungs, and bronchitis are the result of a chill from the bird having been placed in a draught, and differ from each other rather in degree than in kind; the former being a clogging of the minuter structures of the lungs by a sudden rush of blood from the exterior to the interior of the body; and the latter, a similar affection of the larger ramifications of the air passages, or bronchial tubes, which get more or less lined and obstructed by mucus: great warmth is the only cure, as we have already observed, but prevention is easy.

Diarrhœa is generally caused by improper feeding, unless it is symptomatic of consumption or fever; it is treated by a return to a natural diet, and the addition of some powdered chalk to the drinking water, preceded by a dose of castor oil.

Feather eating is a veritable disease, and one, too, that is extremely difficult to cure. Various remedial plans have been suggested, but some cases defy every attempt, and the poor victims remain regular scarecrows to the end of their days, which are generally prematurely ended by cold. Occasionally turning the bird loose into a room fitted with perches and logs of wood will effect a cure; or giving it a companion of its own or a kindred species, though we have known the new arrival to catch the complaint, and soon make itself as great an object as its companion. Fixing a tin collar round the bird's neck, anointing the breast with oil, strewing the bottom of the cage with feathers, have answered in some cases, and failed in others. So that the owner of a feather-eater would do well to give all the above plans a trial, so that if one did not succeed another might. Shower-baths, too, have been suggested, but are not generally successful; parasites must of course be looked for, and guarded against, and if there is any skin irritation, a cooling diet, consisting largely of green food, might be tried: but some cases defy every attempt to cure them, and all are more or less troublesome, requiring a great deal of patience and perseverance if any good is to result from the adoption of remedial measures. Some months since we bought a Green Parrot that had plucked all the feathers off its breast, and was, generally, in very poor condition; we turned it out into a garden aviary well supplied with logs, and the bird is now in perfect plumage, and as sleek and handsome as possible.

If it is desired to breed Parrots, they should be placed in as roomy a cage or aviary as practicable, unless so tame that they can be permitted to have the range of the house; their abode must be fitted up with hollow logs of suitable size, suspended high up against the wall, to keep them out of reach of mice, or a small barrel, with half a cocoa-nut husk firmly cemented to the bottom, may be placed at their disposal, and will often be taken possession of in preference to a hollow log. Boxes with a flat bottom surface are objectionable, on account of the eggs rolling about, and running the risk of taking cold, for few members of the Parrot family make any nest, properly so called, but deposit their eggs on the bare wood of their abode.

W. T. G.

Parrots in Captivity.

Rose-Hill or Rosella Parrakeet.

Psittacus eximius, Russ.
SYNONYMS: *Psittacus capitatus*, Shw., Khl.;
Psittacus omnicolor, Bchst.; *Platycercus eximius*, Vgrs., Gld., etc.
GERMAN: *Der Buntsittich oder die Rosella.*
FRENCH: *La Perruche omnicolore.*

"NOT only is this bird distinguished for the rich colouring of its plumage, but by its lively and active habits, and, in the breeding season, by the comical way it has of singing and dancing": Dr. Russ thus commences his description of the Rosella, which certainly is one of the handsomest of the Australian Parrots, and at the same time one of the most familiar and easily managed of all its congeners: it is about the same size as the Cockatiel, but of rather stouter build, with shorter tarsi and stronger feet and claws: it is found in Tasmania, and all the southern parts of Australia, where it breeds in hollow boughs from October to January, laying from five to seven and even nine eggs at each setting.

The head and neck are bright red, the throat yellow, the cheeks white with a bluish shade, the back greenish olive, the tail green and blue: the back feathers are black edged with green.

The sexes are alike in colouring, but the female is a more sedate personage than her mate, who is vivacity personified; she is a trifle smaller than he is, and the tints of her plumage rather less vivid than his.

The young resemble their parents, but are much less brilliant in appearance: they grow slowly, and seldom assume the adult plumage until quite twelve months old.

The Rosella is a very hardy bird, caring nothing for our severest winters, providing the aviary in which he is placed is sheltered from the north-east winds, and he has some snug box, or hollow log, to which he and his wife can retire during the coldest nights.

In the matter of diet, he will do very well on canary, hemp, and oats, with bread-crumbs, and green food of all kinds added: water he should always have access to, although the authorities at the Zoological Gardens in the Regent's Park are of a different opinion; but in Australia, we have seen flocks of them frequenting the water-holes and creeks at all hours of the day, and so thirsty are they, that they will drink salt-water if they cannot get any other.

The Rosella has very frequently been bred in Germany and in France and Belgium as well as in this country, and may now be almost looked upon as a European bird; the greatest obstacle in the way of its successful rearing being the great resemblance of the sexes, which renders it difficult to secure a pair: the female is rather more subject to egg-binding than the other members of the family which have been bred in our aviaries, and requires to be carefully watched at the commencement of the nesting season, so that she may be placed under treatment at the very first indication of illness.

In the matter of inability to lay her eggs, prevention will be found to be always better than cure, and if the bird be strong and healthy, and has had access to old lime, there will be little fear of her being attacked by this distressing complication: when, however, it manifests itself, the bird must be captured, given a dose of castor-oil, have the vent unointed with the same, and be placed in a warm room: then, when the egg has been laid, she had better not be returned to her mate for some time—not at all if she appears to be in the least drooping.

There is no doubt that during their breeding season in Australia the Rosellas eat a considerable number of insects, notably coleoptera and white ants, which they find in hollow logs and branches: and in captivity we would recommend that a few mealworms, or even scoured gentles be given them when about to nest, as well as when there are young ones to be fed.

It will be readily understood that seven or eight young Parrakeets will consume a large amount of food, and so bread, soaked in *cold* water, must then be placed at the parents' disposal, as well as boiled oats and maize: hay-seed the Rosellas enjoy vastly, and a few handfuls will afford them not only occupation but amusement, and be a wholesome change of food as well.

These birds are susceptible of being tamed, and will even learn to repeat a few words, that is to say the male will, for his lady-wife has not much talent in this direction.

The Rosella is a very noisy bird, almost as much so as the Cockatiel, and is, consequently, scarcely to be recommended as a cage-pet to persons of weak nerves: in a large aviary, however, he will do very well indeed, and constitute one of its chief attractions, for he is, apparently, quite conscious of his personal charms, and is never weary of displaying them to the best advantage: everlastingly in motion, and everlastingly warbling, or chattering, he keeps the whole place in a commotion, and must on no account be trusted with other birds weaker or more defenceless than himself, for, despite his rich dress and gentlemanly deportment, he is a decided "tartar", and, especially during the breeding season, brooks no intruder near his domicile.

In the Australian bush the Rosella is of very common occurrence, frequenting especially the neighbourhood of cultivated lands, where it commits sad havoc with the crops, and is consequently detested by the farmers, who shoot and snare it whenever they get the chance, and destroy its nest without mercy or compunction, which is a stupid thing to do, as these birds always command a good price in the home market, and find ready purchasers in the Australian towns: but the farmer, whether Australian or English, is not a far-seeing personage, and for a slight temporary benefit has no notion but to forego a future emolument that is not very prominently placed before his eyes; and really not always, even then, will he perceive and avail himself of the promised boon.

There is no doubt that a farmer, or anyone else for that matter, who would undertake to keep and breed, not only the commoner, but especially the rarer species of Parrots, whether in Australia or in South Africa, would make a good speculation, and find a ready market for the produce of his aviaries: in this country the climate is too changeable to permit of the success of such an undertaking, unless perhaps in the extreme south, and even there in a warm and sheltered situation only; but in France a *Perrucherie* is by no means an uncommon sight, and is a highly remunerative speculation to boot: Parrots, as a rule, are as easily kept as Pigeons, are very nearly, in fact, considering that the Pigeon has but two young to a nest, and the Parrot has, usually, five or six, quite as prolific, and, with the exception of some of the fancy sorts of the former bird, command a much higher price, and we wonder that Parrots are not more frequently kept on a large scale than they are, for they are excellent eating, and their feathers in much request for ladies' hats and bonnets.

A strong, well-built aviary, plenty of hollow logs, that is all that is needed for a *Perrucherie:* with the exception of the Cockatiels, however, which we have never found to interfere with their fellow-captives, most of the Parrots would require an aviary to themselves, but as the greater number of species are gregarious, several pairs of the same kind may, usually, be kept together, and, providing there is plenty of nesting accommodation about, will not interfere with each other's arrangements: many species, indeed, breeding better in company, than when one pair only is kept.

It is needless to reiterate that a sufficiency of nesting accommodation must be provided, or adieu to peace, and to all hope of increase in the *Perrucherie:* but when this has been attended to the birds will soon settle quietly down, and rarely meddle with one another, for Parrots, on the whole, are sociable birds, and get on better in company than when kept in solitary confinement in a cage; though some misanthropic individuals seem by their conduct to contradict point blank this assertion: nevertheless that there are exceptions to every rule is well known, establishing rather than overturning it: and that this is the case with Parrots, the experience of every aviarist who has kept them in any numbers will, we think, confirm.

When forming a collection of Parrots in an aviary, it will be well to group together the species that more nearly approach each other in size and habits: thus we would not recommend placing Sulphur-crested Cockatoos in the same enclosure with any of the Love-birds, although some species are usually so amiable and accommodating, the Cockatiel for instance, that they will get on in any company, minding their own business with praiseworthy assiduity, without ever inquiring what their neighbours are doing, what they are going to have for dinner, who their relations are, or what means they have for getting on in life, as so frequently happens with the superior creature man.

Needless to plant trees or shrubs in an aviary of Parrots, but hollow logs, trees even, will be a great boon to the inhabitants, affording them not only snug retreats in which to deposit their eggs, and hatch and bring up their young, but also an infinite fund of amusement, not to say delight, and exercise to boot, for nearly all the Parrots are born "whittlers", and if they have not a handy log "convenient", as an Irishman would say, on which to exercise their powerful mandibles, they will find some other and more objectionable mode of whiling away the time, by quarrelling among themselves, or even turning to and plucking out their own feathers by the roots, until they leave themselves quite bare.

As soon as the young of one brood can feed themselves, it is desirable to remove them to other quarters, lest they interfere with the domestic arrangement of their parents, and prejudice the production of another brood. Of course overcrowding must be carefully avoided, and if a separate aviary can be given to each species, so much the better: but this is not absolutely, nor even imperatively necessary, as most of those birds are fond of company, and thrive better in the society of their fellows than when kept alone by themselves.

A male Rosella kept in a cage by himself, especially when he has been brought up by hand, makes a very nice pet, if somewhat noisy, for he gets very tame, and learns to speak fairly well: but a pair are most objectionable, for the male becomes intensely jealous, and spiteful, and his shrieking propensities are quadrupled. A hen Rosella is a very quiet, gentle little bird, seldom or never shrieking, and as she is quite as handsome as her mate, though a trifle smaller, she should be preferred as a cage pet: she is not very intellectual we must admit, but otherwise she is superior to her more demonstrative partner.

Independently of the subject of our next two chapters, there are several varieties of Rosellas to be met with in Australia; for instance, one called by the dealers "the large Rosella" otherwise the Adelaide Parrakeet (*Psittacus platycercus Adelaide*, Gld.), and the small, or Earl of Derby's Parrakeet, which is found in West Australia only; and in addition to these, Gould enumerates other specific varieties, which all bear a strong family likeness to the bird under consideration, and by an unpractised eye are scarcely to be distinguished from it.

Pale-Headed or Mealy Rosella.

Psittacus palliceps, Russ.
Synonyms: *Platycercus palliceps*, Vgrs., Jrdn., Gld., etc.
Platycercus cœlestis, Gr.; *Conurus palliceps*, Cv.
German: *Der blassköpfige Buntsittich.*

THE Mealy Rosella is not, in our opinion, a pretty bird, its pale-coloured head and face give it a decidedly sickly appearance; it is about the same size as the last species, and coloured after much the same fashion, the head and neck of the Rosella being red, and the same parts in its Mealy relation yellow; the cheeks of the latter are white, which tend to increase its really ghastly mien; the upper part of the body is black, each feather being edged with yellow; the under surface is blue; and the lower tail coverts red.

The female resembles the male exactly as to colour, but is, perhaps, a trifle smaller.

The Mealy Rosella replaces in the northern parts of Australia the more vividly coloured bird of the southern portions of the country, so that Dr. Russ is incorrect when he states "*Heimat ein grosser Thiel Australiens.*"

It is not as robust as its more vividly coloured congener, and will not stand the severity of our winters out of doors; it is not very common in the dealers' shops, nor, we believe, very much sought after by amateurs. We cannot encourage our readers to buy it, for the following reasons: it is expensive, rather delicate, at least at first, quarrelsome with its fellows, and, as we have said, presents a sickly, faded, washed-out appearance, that in our eyes is the reverse of attractive.

If any one should decide upon giving it a trial, well and good, feed it on seeds, as the Common Rosella, but give more insect food, such as mealworms, gentles, and black beetles, as well as egg-bread, ants'

eggs, and hemp seed; upon this diet, when once acclimatised, it will do very well, and, in all probability, reproduce its species in captivity, which, indeed, it is stated, *auctore* Dr. Russ, to have done last year in Germany.

It is a strange fact that of two closely allied species, the less brilliantly coloured should be an inhabitant of the warmer region; when, as a rule, the birds of tropical, or sub-tropical latitudes are so much more gorgeously apparelled than their congeners of more temperate climes; for the Mealy Rosella is not, as a glance at the illustrations will show, nearly as brightly coloured as the Rose-hill, which is found in Tasmania as well as in the southern parts of the mainland, and it would be curious to ascertain why this reversal of the usual order should have taken place in the present instance; but the problem is insoluble, unless we suppose that the great heat of Northern Australia blanched the bright colours of the Rosella, and produced the pallid bird which forms the subject of the present notice, which is not a very probable hypothesis: can any one help us to another?

We are inclined to think that the ordinary Rosella, the Pale-headed and the Splendid variety (*Platycercus splendidus*), which must not be confounded with the Splendid Grass Parrakeet (*Euphema splendida*), are not readily distinct species, but rather geographical variations of the same; the first inhabiting the southern, the second the northern, and the third the central portions of Australia; a supposition that might very readily be put to the test by cross-pairing the several birds, and noting whether their offspring were fruitful or otherwise: these experiments, however, are outside the province of most ordinary connoisseurs, and should be undertaken by the Zoological or Acclimatisation Societies, who have the solving of so many interesting ornithological problems in their power, if they would only undertake the task.

The Pale-headed Rosella in captivity is dull and uninteresting when kept by itself in a cage, and is not quite safe to be trusted with Parrakeets smaller than itself in an aviary, so that on the whole it will, perhaps, be as well for the aviarist not to meddle with it at all: nevertheless, as the very difficulty of procuring and keeping a bird is, in itself, a recommendation to some people, we doubt not that purchasers will be found for it, even at the price of £2 10s. now asked for a pair by the London dealers.

Yellow-Rumped Parrakeet,
or Broadtail.

Psittacus flavcolus, **Russ.**
Synonym: *Platycercus flavcolus*, **Gld.**, **Fnsch.**, *etc.*
German: *Der strohgelbe Sittich.*

THE Yellow-rumped Parrakeet is a native of New South Wales, occurring in considerable numbers on the banks of the rivers Lachlan and Darling, but seldom imported into this country, and still less frequently seen on the continent, where it commands a high price, from ninety to one hundred and twenty marks, according to Dr. Russ.

The forehead is crimson, and the cheeks light blue, the crown of the head, back of the neck, rump and upper tail coverts, as well as all the under surface of the body pale yellow. The upper part of the breast is slightly tinged with red; the feathers of the back are black in the centre, and pale yellow on their outer edges; the middle of the wing is pale blue, the spurious wing and the outer web of the basal portion of the primaries are deep violet blue, the remainder of the primaries dark brown; the two central tail feathers are tinted with green at the base, passing into blue towards the tips, the remaining tail feathers have the basal portion of the outer webs deep blue, passing into very pale blue towards their tips, the inner webs are brown, and the extreme tips white; the bill is light horn colour, and the legs and feet dark brown.

The sexes are alike, but the colours of the female are much less brilliant than those of her mate: in size this bird about equals the Rosella, being, if anything, a trifle larger. It is exceedingly rare in this country, and when imported is sold at a very high price, £5 or £6 for the pair, which, considering the frequency with which it occurs

in its native land, is surely rather extraordinary, especially as it is frequently kept as a cage-pet by the colonists, and being a grass-seed feeder is by no means difficult to preserve in captivity.

It was described by Latham, but the engraving accompanying his text is very distorted, and conveys an exceedingly poor idea of the bird, which is really a handsome creature.

In his great work, *The Birds of Australia*, Gould describes this species with his usual felicity, and gives a graphic portrait of it: it is to be regretted that so little is known of this charming Parrakeet in this country, for it is hardy, lively, and beautiful, not too noisy, as Parrakeets go, and susceptible of being rendered very tame; its capacities for the acquisition of knowledge are not very great, but a young male brought up from the nest will learn to repeat a few short words about as well as the Rosella.

In the vast primeval forests of the Great Southern Land the Yellow-rumped Parrakeet is at home, and, of course, seen and heard too, to the best advantage; the glaucous green of the foliage of the *Eucalypti* forms a not inharmonious contrast to the golden yellow shades of his plumage, and his cries pass unnoticed amid the concert of Nature which they contribute to maintain; in confinement, however, he is, as we have said, not inordinately noisy, and may be kept in a parlour without fear of any one being driven out of the house by his cries.

In its native woods this bird breeds in the hollow limbs of trees, a rule to which we know of no exception in Australia, save that of *Pezoporus formosus*, making no nest, but depositing its eggs on the bare wood; it is shy, however, and the nest is difficult to be discovered, and more difficult still to be plundered, when long and careful watching has revealed its site, which is so carefully selected as to be almost inaccessible to man or beast. The breeding season extends from September to February, and there are, usually, two broods in the season, of four or five young ones, which remain in company with the parents until the following spring, when they separate, and each pair sets up housekeeping on its own account.

It is curious to mark the resemblances that approximate the different kinds of Parrots to each other, no less than to consider the distinctions by which they are differentiated, and to observe by what strong evidence it is shown that they must all have been formed upon one original plan, rather than, as the Evolutionists say, have descended, in the lapse of ages, from a common ancestor. That "lapse of ages" is a very handy bridge to get over a difficulty on: but which was the common ancestor? was it a Macaw, or a Madagascar Love-bird? a Grey Parrot, or a Budgerigar?

"The common ancestor has perished, but the links of the chain that connect the whole Parrot family with him remain", it may be replied: well, we prefer another theory, which has the advantage of requiring neither axiom nor postulate to prove its very truth.

The Evolutionist when confronted with a difficulty, meets it by assuming "ages", and saying that this species "has taken to doing so and so", and this other to "doing in such another manner", and "this or that characteristic has been acquired, or relinquished", in virtue, of course, of some inherent power existing in the creature itself to modify its structure. The falsity of this doctrine has been demonstrated over and over again, but it is nevertheless gaining ground; it is the "fashionable" theory of the day, and like all other outcomes of fashion will have its day.

Lories, for instance, are Parrots which "have taken to living among blossoming trees, and feeding off the nectar and pollen of the flowers, instead of seeds and grains. Accordingly, they have acquired a shape well adapted to their new habits, including the short tarsi, and the long filamentous tongue characteristic of these birds", but if anyone should ask when this change of form, this divergence from the original Parrot type took place, he will be told that it all happened long ago, is expected to accept that answer as satisfactory and conclusive, and to ask no more questions, which are embarrassing. "Maman", said a French child to her parent one day, "where is France situated?" "In Europe, my love": "and where is Europe, Maman?" "In the world, my dear": "but where is the world, Maman?" "In the universe, Miss": "but where is that?" *"Tais-toi donc, imbécile, tu m'embêtes!"*

What a distance we have wandered from our Yellow-rumped friends to be sure! France, the world, the universe, and we know not where beside; but not further than the Evolutionist from a true solution of the problem of life. It is a favourite theory with the apostles of the new belief that every one who dissents from their views is a fool; so be it—we prefer our folly to their wisdom.

It is curious, to say the least, that a theory propounded by its author to account for the extinction of a species, should be evoked to explain the origin of species in general by philosophers of quite a different school.

Professor Owen thus writes in reference to the *Origin of Species*, by the late Mr. Darwin:—"The influence of the contest for existence, amidst the changes of the circumstances to which an animal has been adapted, on the extinction of species, was first propounded by the author, in his fourth memoir on *Dinornis*, 1850, (*Trans. of the Zool. Society*, vol. iv., p. 15.) The same principle has since been evoked to explain not only the extinction but the origin of species. Mr. Wallace (*Proceed. Linn. Society*, August, 1858, p. 57,) assumes that a variety may arise in a wild species, adapting it to changes in surrounding conditions, under which it has a better chance of existence than the type-form from which it deviated, and of which it would take the place. Mr. C. Darwin had, previously to Mr. Wallace, conceived the same application of this principle, which he illustrates in his work *On the Origin of Species*, by many ingenious suppositions, such as the following:—'To give an imaginary example from changes in progress on an island, let the organization of a canine animal which preyed chiefly on rabbits, but sometimes on hares, become slightly plastic; let these same changes cause the number of rabbits very slowly to decrease, and the number of hares to increase; the effect of this would be that the fox, or dog, would be driven to try to catch more hares; his organization, however, being slightly plastic, those individuals with the lightest forms, longest limbs, and best eye-sight, let the difference be ever so small, would be slightly favoured, and would tend to live longer, and to survive during that time of the year when food was scarcest; they would also rear more young, which would tend to inherit these slight peculiarities. The less fleet ones would be rigidly destroyed. I can see no more reason to doubt that these causes in a thousand generations would produce a marked effect, and adapt the form of the fox or dog to the catching of hares instead of rabbits, than that greyhounds can be improved by selection and careful breeding.' Yet this condition of things, if followed out to its full consequences, seems to lead only to my original inference, namely, an extinction of species; for, when the hares were all destroyed the long-legged dogs would perish. At most there could but be a reversion to the first form and conditions. For, as the hares decreased in number, that of the rabbits would increase; the changes of organization that fitted the dogs for catching hares being such as would detract from their power of unearthing rabbits. A variety with a shorter and stronger foot might arise, and would be the first to profit by the preponderance of the burrowing rodents. The individual dogs with the strongest and shortest limbs, let the difference be ever so small, would be slightly favoured, live longer, rear more young inheriting the rabbit-catching peculiarities; the less fossorial varieties would be rigidly destroyed, etc. It is an argument in a circle."—(Owen: *Palæontology*, p. 435.)

Exactly so: and if in the foregoing somewhat lengthy extract we read Parrot instead of 'dog', and blossoms and seeds instead of 'hares and rabbits', we have the case of our *Psittacidæ* to a nicety, and equally untenable, the only real explanation of the origin of species being, as Professor Owen says: the existence of "a continuously operative secondary creational power", which even the late Mr. Darwin admitted in the following terms:—"Certain elemental atoms had been commanded suddenly to flash into living tissues", but limits the number of original progenitors to four or five: while "Analogy", he adds, "would lead me one step further, namely, to the belief that all animals and plants have descended from some one primordial form, into which life was first breathed." (p. 414.) Surely a most unnecessary hypothesis.

"Observation", continues Professor Owen, "of the actual change of any one species into another, through any or all of the above hypothetical transmuting influences, has not yet been recorded; and past experience of the chance aims of human fancy, unchecked and unguided by observed facts, shows how widely they have ever glanced away from the gold centre of truth."

That man has the power of producing and perpetuating varieties in many species of domesticated animals is undoubted, we need only point to the many breeds of dogs, pigeons, and poultry; but these varieties are not distinct species, they breed *inter se*, and the progeny is fruitful; while the offspring of species that bear a much closer resemblance to each other than a pouter-pigeon, for instance, does to a fantail, are barren, because the parents belong to different species and are not descended from one another, witness the case of the horse and the ass; but the columbine cross to which we have referred is capable of reproduction, because the parents, though having little likeness to each other, are varieties of the one species, and not at all distinct: man cannot make species, though he can produce varieties, neither can "circumstances" or "changes of condition"; to create is the prerogative of the Most High, whose works are inimitable.

To return to our Yellow-rumped Parrakeets: they are nice, quiet, gentle birds, susceptible of being perfectly tamed, are easily fed and kept in captivity, but are by no means descended from either a Great Vaza Parrot, or a Black Cockatoo.

New Zealand Parrakeet

Psittacus Novæ Zelandiæ, Russ.
Synonyms: *Cyanoramphus Novæ Zelandiæ*, Snc.;
Cyanoramphus Novæ Guineæ, Bp.; *Cyanoramphus auclandicus*, Bp.;
Euphema Novæ Zeelandiæ, Mus. B. P.;
Lathamus Sparrmanni, Lss.; *Psittacus pacificus*, Reyneri, Cooki, Gr.;
Platycercus pacificus, Vors. German: *Der Ziegensittich*.

NEW ZEALAND contains a number of ornithological curiosities among its peculiar fauna, and although the subject of the present notice can scarcely be classed among them, it is still a sufficiently remarkable bird to merit a little attention, which, indeed, it has already received at the hand of writers on the subject of Natural History, as witness the formidable array of scientific names bestowed upon it, to which we might have added several more, but that we considered it quite sufficiently burdened as it is.

As regards plumage, *Cyanoramphus* presents nothing very striking to attract the observer's notice, while its capacity for acquiring knowledge is not excessive; its disposition, if shy, painfully so in point of fact, is decidedly gentle and pleasing, and it has a rather agreeable voice, which it can modulate into a passable imitation of articulate sounds; its own wild notes, however, are sufficiently agreeable, and it can sing a peculiar kind of song, which has been likened by its German admirers to the notes of a hurdy-gurdy of superior construction, much modified by distance, as well as to the vocalization, under similar conditions, of our ancient acquaintance Punch of puppet-show notoriety; whence the names of *Kladderaduts-Sittich* bestowed upon it by its Teutonic friends, in addition to the appellation of *Ziegensittich*, by which it is more generally known in Germany.

When first imported these birds are decidedly delicate, and numbers of them die before they become acclimatised, but once they have recovered from the effects of the voyage, and have become accustomed to their new food and surroundings, they got on quite as well as any of the Australian Parrots, Parrakeets, or Lories, with which we are acquainted.

Dark green is the prevailing colour of the plumage in these birds, but the forehead, cheeks and rump are red; the flight feathers of the wings are blue, and a few slight shades of the same colour appear in different parts of the body, especially in certain lights.

The female is rather larger than the male, which is about the size of the Rosella, and the red in her plumage is of a duller shade than in the case of her mate.

As its English name implies, this bird is a native of New Zealand: it passes a good deal of time upon the ground, hence its legs are long, the wings short, and the feet have no great prehensile power, although the bird makes good use of one of them, usually the left, for holding up its food: the beak, too, is slight, and better adapted for eating grass seeds, than maize or hard corn.

The nest is made in a hollow branch, and four or five white eggs are laid on the bare wood, and hatched in about eighteen days: a recent number of Dr. Russ's *Die gefiederte Welt* contained an account of a brood of these birds, that had been successfully reared in Germany; the first instance of their reproductiveness in Europe with which we are acquainted: they are reported to be insectivorous in their habits, but this we doubt; none of the specimens we have seen would touch anything of the kind, but lived chiefly on canary seed, which they were fond of scratching up with their feet at the bottom of the cage in which they were kept.

They are very gentle birds, and well deserve the epithet *pacificus*, given to them by Latham, Vieillot, and other writers. They are not able even to protect themselves from their stronger-billed congeners, and should not be enclosed with any of the *Platycerci*, Cockatoos or larger Parrots; on the other hand, they will not molest the tiniest Finches, which evince no fear of them, but, emboldened by the gentleness of their habits, scruple not to play all kinds of tricks upon them, which the New Zealanders treat with good-humoured contempt: thus when one of these birds approaches the seed-pan, where a Canary, Java Sparrow, Silver-beak, or even an Avadavat is occupied in eating, and the latter opens its beak and scolds, as the manner of these birds is to do, *Cyanoramphus* turns him about, though never so hungry, and patiently waits to satisfy his appetite until the coast is clear.

So gentle, peaceable, and loveable a bird ought to be a general favourite, and is with every one who knows him, so that we expect before long to find him one of the commonest of cage pets.

As we have said, his voice is peculiar, but not disagreeable, resembling somewhat the subdued bleating of a young goat, whence his German name of *Ziegensittich*. The female is a very silent bird, and beyond a faint squeak, or bleat, now and then, we have not heard her make any kind of noise.

When first imported these birds should be fed on oats, part of which should be boiled before being given to them: many of them will also eat crumb of bread steeped for a few minutes in cold water, and then squeezed nearly dry; they drink a good deal, and should always have access to a free supply of clean water.

Green food must be given sparingly at first, as they are apt to eat it so greedily as to bring on diarrhœa: after a time they can be accustomed to canary seed as their general diet, adding oats and hemp seed now and then as a variety.

Sometimes they are imported in large numbers, and can then be bought cheaply; at other times they are scarce, and barely to be had at any price.

The Hon. and Rev. F. G. Dutton's account of the New Zealand Parrakeet (Cyanoramphus Novæ Zealandiæ).

THIS bird resembles the Pennant's Parrakeet in its habits: it is shy and gentle, a quiet bird, and very fond of bathing. I have known three, of which two were talkers, but none of them tame enough to allow themselves to be handled. The two that talked said several sentences, but did not pick up anything fresh. They are very attractive cage-birds for any one who likes quiet birds, but they are rather wanting in character. I dare say they would breed readily in an aviary, and then, if any one would be at the pains to rear the young by hand, I think they would make the most delightful pets. Mine eat hemp and canary seed, and had plain water.

GOLDEN-CROWNED PARRAKEET OF NEW ZEALAND.

Psittacus auriceps, Russ.
SYNONYMS: *Cyanoramphus auriceps*, Bp.; *Cyanoramphus Malherbi*, Sng.; *Psittacus pacificus*, var. d., Gml.; *Euphema auriceps*, Lchtst.; *Platycercus auriceps et Malherbi*, Gr. GERMAN: *Der Springsittich.*

THIS charming bird, one of the most delightful of all the Parrot family, according to Dr. Russ, is, as its English name denotes, a native of New Zealand, as well as the species described in the preceding chapter, and is about the same size as the Turquoisine, but of heavier and less elegant build than the latter bird: the plumage on the upper parts of the body is a dark grass green, with here and there a shade of blue; on the breast and abdomen the feathers are of a lighter, or rather a yellower, shade of green than on the back and wings, the front edges of the latter are blue: the tail is dark green above, and grey underneath.

The legs are long and of a greyish slate colour, the claws black, and of unusual length: the upper mandible is blue with a black tip, and the lower horn colour. Surrounding the beak is a narrow line of an intensely red colour, which seems to be continued into the eye, the iris of which is a brilliant red: the top of the head is pale golden yellow.

The long shanks of this bird enable it to run and hop with great freedom, and have obtained for their possessor the German name of *Springsittich*: it has a curious habit, which seems to be peculiar, and common to the Parrots of New Zealand, of scratching with its feet in the sand and dust after the manner of the *Gallinaceæ*, and it also makes use of them to hold its food, after the well-known fashion of the Grey and other large Parrots.

GOLDEN-CROWNED PARRAKEET.

Although the beak of this Parrakeet is very long and sharp, and its claws of unusual size and strength, it is a remarkably gentle and timid creature, never bites, even when taken in the hand, or makes itself in any way a nuisance to its owners or to its companions in the aviary, where it permits itself to be bullied by birds not a third or quarter of its size, without evincing the least resentment.

The voice of the Golden-crowned Parrakeet is soft and musical, and we have never heard it scream. It becomes exceedingly tame in captivity, to the extent even of accompanying its owner out of doors; it is easily frightened, however, and as its wings are strong, and its flight powerful, it is apt to stray away when terrified, but invariably returns when the alarm is over.

The female bears a strong resemblance to her mate, and can only be distinguished from him by comparison, when she will be found to be a trifle smaller, and to have a rather paler frontlet than he.

Although so tame and gentle, there is no record of these birds having, so far, bred in captivity; yet they are hardy enough to pass the winter in an unheated bird-room without inconvenience to themselves, and it is probable that in a well-sheltered aviary they might be even safely wintered out of doors.

Occasionally they are imported in large numbers, and may then be bought cheaply; as a rule, however, the dealers charge two or three pounds a pair for them.

The Golden-crowned Parrakeet requires to be fed and treated as recommended for its red-fronted congener, described in the preceding chapter, and seems to be particularly fond of bread-crumb soaked in cold water, which, probably, reminds it of the soft seeds upon which it was accustomed to feed in its native land; it drinks freely, and should never be without water, both for bathing and drinking purposes.

It is a very cautious bird, and though extremely fond of a dip, displays a considerable amount of sagacity in the indulgence of its propensities, and will by no means rush inconsiderately into danger: we lately witnessed one of these birds anxious to take a bath, but fearful of the depth of water, make a circuit round the pan, as if considering what steps to take, and then, holding on tightly by one foot to the edge of the vessel, back cautiously into the water to about a third of its own length, when, thinking, doubtless, it had gone far enough, it flapped its wings vigorously, wetting itself completely in a few seconds, and then, drawing itself up carefully out of the pan, flew off to a perch to dry and arrange its feathers.

This bird is excessively fond of washing itself, and also drinks freely, notwithstanding which facts we have seen the poor creature condemned to an agony of chronic thirst in a dry and choky atmosphere, where, we were not surprised to hear, they seldom survived for any length of time: we remonstrated with the person in charge, but without avail, Parrots never drank, he said, and we found that it was useless to attempt to convince him of his error, in which his superiors of course participate.

Superstitions die hard, as a rule, but are killable nevertheless, and it will be no fault of ours if this "vulgar error" is not decently buried out of sight ere long.

In refutation of the notion that the Golden-crowned Parrot of New Zealand does not long survive in captivity, we may mention that we have one that has lived over two years in confinement, and appears to be quite healthy and contented, and we are not without hopes that he and his mate may be induced to breed in their aviary next season.

Although an undeniably charming bird, we consider that fifty shillings, the price quoted in a dealer's list now lying before us, is quite too long a figure to pay for a comparatively common bird, which we have by no means found difficult to preserve, whether in cage in doors, or garden aviary: during the moulting season, however, it is wise to take it into the house, as the evenings are chilly in autumn, and we found that the process of renewing its feathers was, under such circumstances, protracted; and that it was better to take the poor bird into the house.

If kept in a cage this Parrakeet becomes tolerably familiar, and we have no doubt would learn to speak, at least a few words, but we have not tried to teach ours, which, we must say, are not particularly tame; but then, as we have said, we have not in the least attended to their education. Occasionally these birds can be picked up cheaply, and amateurs should be on the look out for such opportunities, and make their purchases as soon as possible after the birds have reached the dealers' hands; they may lose one or two, it is true; but, on the whole, we think they will be thus more likely to secure healthy subjects, than if they waited for several weeks, by which time the high temperature at which foreign birds are chiefly kept in the shops of the dealers cannot fail to have had an injurious effect upon their constitutions.

This Parrakeet is very slim of figure and can squeeze himself through a very small opening without the slightest difficulty, so that in placing him in a cage, care must be taken that the bars are sufficiently close together to prevent him taking French leave. The first bird of this species we possessed was, upon its arrival, transferred to an ordinary Parrot cage in which it seemed to make itself very much at home: but scarcely was our back turned when we heard a terrible commotion, and, to our consternation, behold our new acquisition in the mouth of our big black tom-cat. Naturally we thought the poor stranger's doom was sealed; and to seize the cat, choke him till he dropped the Parrakeet, and prepare to bind up the wounds of the latter, if indeed he were yet alive, was the work of but a few seconds, when to our surprise the liberated *Auriceps* flew off as if nothing had been the matter, and darted round and round the room in evident enjoyment of his recovered liberty: nor was it without some difficulty that he was eventually recaptured.

When at length we had secured our truant, we found the poor fellow to be so seriously wounded on the back, between the wings, that we gave him up for lost; we put him in a small cage, however, he had walked out between the bars of the large one, as one might do through an ordinary door-way, and in a few days Richard was himself again, whereupon we turned him into a large aviary out of doors in company with a mixed collection of foreign birds, where he soon made himself at home.

We have kept a number of different kinds of Parrots and Parrakeets in our time, and the subject of the present notice stands almost as high as any of them in our estimation.

Since writing the above, we have been obliged to remove our favourite in-doors: about the beginning of October he began to moult, and appeared to feel the sharp weather that soon afterwards set in so keenly, that we thought it would be decidedly cruel to subject him to it any longer, but another change of temperature taking place, we decided to leave him where he was, for at least a few days longer. One morning, however, when we entered the aviary to feed the birds, *Auriceps* was nowhere to be seen!

High and low, in every box and husk we looked for him, in vain: he was gone, and the mysterious part of the matter was that we could see no possible way by which he could have made his escape. He had not been carried off by a marauding rat, or dragged through the wires by a prowling cat, for, in either case, some of his feathers would certainly have been lying about. What could have become of him?

In one corner of the aviary, on the ground, stood an old cage, we lifted it up and crouching under it, in a burrow which he had evidently excavated for himself in the soft earth, lay *Auriceps, perdu!* But the cold ground had chilled him, he was cramped and unable to fly: we thought he had taken his death, as the saying is, and removed him at once indoors; but in a few days he was himself again; so after keeping him for a while in an old canary-breeding cage, we turned him, not without some misgiving, into a large aviary-cage inhabited by a couple of Blue-winged Parrakeets, which dashed wildly about, apparently much afraid of the intruder.

The Golden-crowned one, however, took no notice of them whatever, and in the course of a few minutes the tiny couple, having satisfied themselves that the new arrival harboured no sinister designs against their peace, settled quietly down again, and recommenced their interrupted combing of each other's heads, as if nothing had happened, and there he has remained ever since, looking as fresh and comfortable, as if he had just come in from a ramble through those far-off New Zealand woods, which he is fated never to see again.

In a recent number of *Die gefiederte Welt* there was an account of a brood of Golden-crowned Parrakeets that had been successfully reared in Germany.

BLUE BONNET PARRAKEET.

Psittacus hæmatogaster, RUSS.
SYNONYMS: *Psephotus hæmatogaster*, GLD.;
Platycercus hæmatogaster, FNSCH.; *Euphema hæmatogaster*.
GERMAN: *Der Blutbauchsittich*.

THIS handsome bird is a native of New South Wales, but is not very frequently seen there, and is of comparatively rare occurrence in England, and rarer still on the continent. "*Einer der allerseltensten*", says Dr. Russ, who quotes its price at ninety marks, "*und darüber fur die Prch.*": that is, £4 10s. and upwards for the pair.

The greyish white beak is surrounded by a blue mask, of a deeper shade on the forehead than on the throat; the back, wing coverts and breast are yellowish grey, the sides of the wings and the tail blue; the sides and under tail coverts, yellow, and the abdomen bright red.

The female is coloured like the male, but the shades of her dress are duller than his.

Some diversity of opinion exists as to the merits and qualifications of this bird as a cage pet: Mr. Wiener writing of it says: "I do not know a more pleasing Parrakeet than the Blue Bonnet, whom I used to consider the clown of a collection of Parrakeets I had at one time. A pair of these birds used to play together like kittens, rolling over and over in the sand, or sitting on the perches and cawing to each other in the most amusing manner. Whenever their cage was fresh sanded, they picked out all the small stones, and cleverly arranged them in a row on a ledge in their cage. This rare Parrakeet is probably one of the most intelligent of all the Australian Parrots, although I have not heard that any have learned to talk, nor do any appear to have been bred."

Mr. A. Johnson, of St. Olave's Grammar School, writing in *The Bazaar* of 21st. March, says, "These birds (Blue Bonnets), which Dr. Russ describes as among the rarest imported, have of late been seen rather more frequently. They are certainly the hardiest of the Parrot tribe in captivity, not excepting the Budgerigar. Introduced into an exposed out-door aviary last spring, immediately after importation, without any attempt at acclimatisation, they have undergone hardships, both as to exposure and food, under which even the Cardinal has succumbed, and yet they never had an hour's sickness. They are seen to the best advantage when seated on a lofty perch, with their primrose underside, so curiously aproned with blood-red, exposed to view; their elaborate bowings and antics are calculated to produce shouts of merriment. They seem the mildest of the inmates of the aviary, but they are really its most insiduous assassins. I have found young birds with their pinions cruelly mutilated, although they were apparently safe in small cages; young Budgerigars, valuable Bourke's Parrakeets, Turquoisines, and others, dead or dying, with their wing joints mutilated, or their heads smashed; and I never was able to trace the assassins, until one day I saw my innocent looking pets, sidle up to a delicate graceful Dove, seize him by the wing, and begin to gnaw him savagely. They will live for months with smaller birds on the most friendly terms, but in the end they will clear an aviary of all weaker than themselves, although like true assassins, they never attack one of their own size. It is only fair to say that these are only imported birds, and that some I have bred myself have not developed this murderous tendency. For hardiness, intelligence, grace, and most amusing ways, commend me above all to the Blue Bonnet, but be sure to keep him with birds who are his match in strength, or, better still, in a small compartment by himself, when he will be a model of good behaviour."

Having no one to fight with, or to murder, he will be perfectly inoffensive, no doubt; but as we have already remarked more than once, birds vary in their dispositions, as Mr. Johnson himself admits, and one pair of Blue Bonnets will be found to be as peaceable and orderly, as another is cantankerous and objectionable.

In size these birds are somewhat less that the Cockatiel, but of more slender build. As Mr. Johnson remarks, they are very hardy, and thrive exceedingly on a diet of oats, to which they are especially partial, canary seed, millet, hemp, and boiled maize: they are very fond of green food of all kinds, and especially of the bough of some tree, such as elm or poplar, which they soon peck to pieces with every manifestation of delight.

Much of the mischief wrought in aviaries by one sort of bird, or another, is due to overcrowding: better keep four birds comfortably, than a dozen where they have not room to turn round without treading on each other's heels.

Blood- or Red-Rumped Parrakeet.

Psittacus hæmatonotus, Russ.
SYNONYMS: *Psephotus hæmatonotus*, GLD., BP., etc.; *Platycercus hæmatonotus*, WGM., GR., etc. *Euphema hæmatonotus*, MUS.
GERMAN: *Der Singsittich.*

THIS extremely pretty and elegant Parrakeet, also known by the name of Red-back, is rather less than the Cockatiel in size: its general disposition, at least as far as our experience of the species goes, is exceedingly unamiable, and we cannot recommend its being kept with other Parrakeets: a pair, however, placed in a roomy aviary by themselves, will very soon set about reproducing their species, and succeed to admiration, which, as the bird is handsome, extremely lively, hardy, and gifted with quite a musical voice, is a fact to be remembered.

The general colour of the plumage in the male is rich grass green, with a blue reflection in certain lights, especially on the head and face, the belly is yellow, the rump red, the shoulders blue, and the tail dark bluish green; the colour of the beak is dark horn, the legs and feet grey.

The female is greyish green with a mottled appearance, arising from the fact of each feather being margined with a narrow line of a deeper shade of the general colour of the plumage, the shoulders are blue, the tail has a deep shade of blue, and the rump is bright green. So dissimilar are the sexes in appearance that they have been taken for different species by some of the earlier writers on Australian Parrakeets.

These birds breed as freely in captivity as the Budgerigar or the Cockatiel, laying from three to five small eggs which the female alone incubates, her mate rendering her no assistance, his cheerful song, as he sits at no great distance from the hollow log that contains the precious eggs, excepted; for he does not even feed her, nor, as far as we have been able to ascertain, does he feed the young until these have left their natal log, and are able to fly about after him, and importune him for food.

We have found that half a cocoa-nut husk cemented into a small box made a capital nest that was much appreciated by these birds, which do not seem to care about excavating a dwelling for themselves, when a ready-made one has been placed at their disposal.

We fed on seeds only, canary, millet, hemp, oats, of which they were particularly fond, and dry bread-crumb: Dr. Russ, however, recommends the following diet when the birds are nesting:—"Egg-bread, ants' eggs, softened rice and fruit; also mealworms, green food, and poppy seed."

We cannot endorse his further statement that they are sociable with little birds, "*Verträglich unter kleinen Vögeln*", but they do nest readily (*leicht*), and bring up three or four broods in the season, as the doctor further relates: they are hardy, too, and will pass the coldest and most severe of our winter out of doors without injury.

The male and female are very much attached to each other, so much so that if one of them should escape, it will, after a fly round, return to its companion, and suffer itself to be captured without resistance.

These birds are very strong on the wing, and it is quite a pretty sight to see them wheeling round and round in the sunshine, or darting in and out among the trees, with the foliage of which their feathers harmonize so well in colour.

We believe that, like most of the Australian Parrots, the Redrumps are partially insectivorous, but they will, nevertheless, thrive perfectly well without insect food. In winter it is advisable to give them plenty of hemp, and they will then touch little else but that valuable and highly nitrogenized diet.

It is a pity they are so tyrannical and quarrelsome, for otherwise they are very nice, and the song of the male bird, especially during the season of love and courtship, is, as Mr. Wiener says, "quite surprisingly agreeable."

Dr. Bodinus, of Cologne, was the first person who bred these birds in Europe, but since then they have bred in innumerable aviaries in this country, as well as on the continent; and in fact more Redrumps are now yearly bred in Europe than are imported from Australia, and the price has fallen to about twenty or twenty-five shillings a pair.

The young resemble their parents in a general way when they leave the nest, but their colours are duller and fainter in shade than those of the old ones, from whom, in about six or eight months, it is impossible to distinguish them.

We have read of hybrids between the Redrump and the Rosella, as well as several other kinds of Parrakeets, and we are quite prepared to believe in the possibility of such a cross, or crosses, for a female of this species that was in our possession for a considerable time actually paired with a Madagascar Love-bird (*Agapornis cana*), and, had she not fallen ill, would doubtless have bred mules with him.

Although, as we have said, these birds are hardy, the females are often troubled with egg-binding, and as this complication is of decidedly more frequent occurrence in aviary-bred than in imported specimens, we incline to the belief that debility is the cause of the misfortune, and that none but thoroughly strong and healthy birds should ever be put up for breeding; inattention to this simple rule entailing much loss and disappointment on the amateur, as well as being the cause of much suffering, and of death to the hapless bird herself.

We once had a fine healthy-looking hen Redrump that never laid an egg larger than that of a Budgerigar; and strange to say, these miniature productions were devoid of yolk, and consequently sterile; she was aviary-bred, and there had, probably, been a good deal of in-breeding in her family.

In-breeding, as bird-fanciers know, is soon productive, in most cases, of disastrous consequences, and should always be avoided, unless it be desired to perpetuate some accidental peculiarity, or "sport;" in which case the offspring will, after a few generations, cease to breed among themselves, and the new variety die out, unless the strain be re-invigorated by the careful introduction of new blood.

"There are no song birds in Australia" is a complaint, more or less founded on fact, one often hears; but the Redrump sings, actually sings a very passable song, a fact which has procured for him in Germany the name of *Singsittish*.

We have not met with a talking Redrump, but as they can be rendered very tame and confiding, it is quite as likely that a young male, brought up from the nest, would learn to speak, as many other varieties of Parrakeets, including the Budgerigar, have done. It is, however, astonishing how greatly these birds differ among themselves in disposition and character, which accounts for the fact of their being described by one writer as gentle, tame, and confiding, and by another as irreclaimable and wild.

MANY-COLOURED PARRAKEET.

Psittacus multicolor, Russ, Khl., Tmm.
Synonyms: *Platycercus multicolor*, Vgrs., Wgl., Fnsch.;
Psephotus multicolor, Gld., Bp., Gr.; *Euphema multicolor*. German:
Der vielfarbige Sittich.

A VERY beautiful, but most unsatisfactory bird is the Many-coloured Parrakeet: a glance at the engraving will convince the reader of the correctness of the former assertion, and our word may be taken in confirmation of the second; for we are by no means alone in our opinions. "A glance at the illustration", writes Mr. Wiener, "will convince the reader that the Many-coloured Parrakeet is one of the most beautiful birds of his tribe. The female in my possession happens to have endured for years, but I feel sure that some day she will be so unreasonable as to die, without any palpable reason, as several of her mates did long ago."

"*Lebhaft und anmuthig, ebenso liebenswurdig als prächtig*", writes Dr. Russ, "*doch kommt es vor, das auch er plötzlich ohne ergründbare Ursache erkrankt und stirbt.*" (As lively and charming, as it is desirable and beautiful, it nevertheless happens that it will suddenly and without accountable reason fall sick and die."

Another writer speaks of these birds as if he had found them hardy, but as he also says that the female is all but indistinguishable from the female Redrump (*Psittacus hæmatonotus*,) his experience of the species was probably limited.

As Dr. Russ says, the male is a handsome fellow, the general colour of his plumage is rich emerald green, darker above than on the lower surface of the body, the forehead is yellow, the crown of his head deep crimson, the shoulders and the sides yellow, the flight feathers and the longest tail feathers deep greenish blue, the abdomen and the thighs blood red.

Many-Coloured Parrakeet

The female, as usually happens among birds, is much more soberly attired: her forehead is yellow, but of a paler shade than in the male, the top of her head green, the back and secondaries of the wings grey, the neck and breast reddish grey, the primaries green with black extremities, the abdomen yellowish green, the under tail coverts yellow, the tail greenish blue, but lighter than in the male, and her shoulder patch, instead of being yellow is red; so that she bears a considerable resemblance, not to the female Redrump, but to the female of the Beautiful Parrakeet (*Psittacus vel Psephotus Pulcherrimus*), from which however she can be distinguished by the larger extent of her red shoulder patch, while from the young male of the latter species she will be known by her breast of reddish brown, and the deeper colour of her wings and tail.

Not very numerous in their native country, these birds are not frequently imported; but when they do arrive, they are readily sold to amateurs at a high figure, notwithstanding the fact that they seldom endure for any length of time in captivity, for they and the following species, called the Paradise Parrakeet, are really Lories, and during the greater part of the year feed on the pollen and nectar of the *Eucalypti* and other flowering trees of their native land, for which sponge-cake is at the best but a poor substitute.

During the winter there is no doubt that these birds subsist on seeds, but those are always soft, and to keep them alive in this country nature should be imitated for them as nearly as can be: thus their millet and canary seed must be soaked in cold water for some hours, and then left to drain before been given to them; soft sponge-cake and bruised figs must also be supplied, and in summer in addition to the above, they should have an abundance of groundsel tops, cabbage or brocoli flowers, mignonette, dandelion flowers, and so on, and especially the blossoms of the lime-tree: nor should half a dozen mealworms *per diem* be omitted for each bird.

Attention to these rules will enable the amateur to keep these beautiful birds successfully, and doubtless to breed them too; but with every care they are apt to look a little puffy one day, to be found the next morning by their disconsolate owner dead; the cause, constipation, a flow of blood to the head, or a rapture of a blood-vessel in the brain: or, pining for their natural food, they sometimes fall into a decline and gradually fade away, though their usual exit from this troublesome world is painfully sudden and unexpected.

The Many-coloured Parrakeets are very gentle and inoffensive creatures, never interfering with the other inmates of the aviary, nor even attempting to defend themselves when attacked, so that care must be taken not to place them with any of the strong-billed and mischievous varieties of their compatriots; they may be rendered very tame by the judicious administration of their favourite morsels, to obtain which they soon overcome their natural timidity and fear of man.

In their native country they breed in the hollow branches of trees, laying three or four white eggs on the bare wood; and have usually two broods during the season. We have no authentic record of their having been bred in this country, or even on the continent; for although some were advertised recently as "aviary-bred", we cannot believe that the birds so offered ever saw the light in this changeable clime.

It is a pity they are so delicate and hard to preserve, for, with one or two exceptions, they are the most beautiful and desirable of all the Australian Parrakeets. When in good health they are very lively and active, and the male has a soft and musical voice, of which, especially during the pairing season, he avails himself pretty freely.

As might be expected from their gentle and inoffensive disposition, the sexes are strongly attached to each other, and are really much more truly deserving of the appellation of Love-birds than the short squat little creatures upon whom it is usually bestowed.

We should take it as the greatest of favours if those readers who may attempt to keep these beautiful birds as recommended by us, or upon any other plan, would communicate the result of their endeavours to us through our publishers, for it is only by such interchange of experiences that we can ever hope to arrive at a solution of the difficulty hitherto experienced by amateurs in preserving this and the species that forms the subject of the following chapter.

When in health, and gradually weaned off to seed, sponge-cake and bruised figs, not forgetting the mealworms, the Many-coloured Parrakeets are not particularly susceptible to cold, but during severe weather avail themselves of the cozy shelter of a hollow log: in fact they get on much better without, than with, artificial heat in winter, and we have seen them successfully wintered out of doors, during the severe seasons of four and five years ago, and never saw finer, or, apparently, more healthy and vigorous birds.

M. Alfred Rousse, of Fontenay-le-Comte, records a case of successful reproduction of the Many-coloured Parrakeet last year.

Beautiful or Paradise Parrakeet.

Psitiacus pulcherrimus, Russ.
Synonyms: *Psephotus pulcherrimus*, Gld., Bp.; *Platycercus pulcherrimus*, Gr., Fnsch.; *Euphema pulcherrima*. German: *Der Paradiessittich*.

MORE lovely, if possible, than the Many-coloured Parrakeet, the subject of the present chapter is even more unsatisfactory as a pet.

No one can see it without desiring to possess so beautiful and graceful a bird, and large sums are constantly being paid for handsome specimens by amateurs: but, alas! one in a thousand survives a few months, and—dies suddenly in a fit one day.

Much acrimonious controversy has been expended on the subject of the endurance or non-endurance of these birds in captivity. Dr. Russ and Mr. Wiener consider that it is all but impossible to preserve it for more than a few months, while other writers look upon it as not more difficult to keep than a Budgerigar; one author calls it a "Grass Parrakeet", but admits that it "requires skilful management", as without "the most watchful care", it has "a nasty habit of shuffling off this mortal coil without giving any previous intimation of its intention so to do."

Such also is our own experience, but we have nevertheless seen specimens in magnificent plumage that survived in an out-door aviary for nearly two years, and may be there yet for anything we know to the contrary; they were fed and treated as we have recommended in the last chapter for the Many-coloured Parrakeet: and in Germany eggs have been produced, but, as yet, no young of this species have been reared in captivity; at least, to our knowledge.

Twelve or thirteen inches in length, of which the tail occupies five or six, these slim and elegantly-shaped birds are natives of New South Wales, where they feed on the honey and pollen of flowers, flies and small insects, and in winter on such insects and seeds as they can find.

The sexes differ immensely in colour, the male is gorgeously apparelled, and the female as soberly clad as her mate is gay. The top of the head of the male is dark grey, the back and wings are of the same colour but a shade lighter, the forehead is bright red, the face, neck, and breast a wonderful combination of blue and green, so blended that in one light the one colour preponderates, and the other in another; the rump is red, flecked with yellowish white spots, the tail is green, shading off to blue at the extremities of the feathers, a band of scarlet marks the shoulders, the beak is grey, and the feet and legs pale slate colour.

The female is yellow in those parts where her mate is green and blue, and pale green where he is yellow, her head and wings are of a paler grey than the male's, and her shoulder bands are yellow with a tinge of red, a few specks of the same colour appearing on her breast.

The young males can be distinguished from their mother, by their red frontlet, red shoulder stripes, green cheeks, and reddish abdomen, while their wings and back are nearly as dark as those of their father.

It is a pity these beautiful creatures are so difficult to keep, for, apart from their beauty, there are few foreign birds more amiable and inoffensive in their habits, or more susceptible of being completely tamed; and if only a suitable diet could at all times be devised for them they would be more frequently met with in the aviaries of amateurs than is at present the case.

Although generally classed by writers with the Grass Parrakeets, *Euphemæ*, or with the *Psephoti*, the Beautiful is more nearly related to the *Trichoglossi*, and if this fact be borne in mind, and its treatment assimilated as much as possible to that recommended for the former birds, as well as for the Many-coloured Parrakeet, a considerable advance toward a solution of the difficult problem of how to preserve them in captivity will have been made.

Although an enterprising breeder recently advertised aviary-bred specimens of the Beautiful or Paradise Parrakeet for sale, we venture to doubt the fact of their having been bred in this country, or even on the continent of Europe: it is just possible they may have been reared at the Cape of Good Hope, where Blue Mountain and other Lories have, we know, been produced in confinement; but flowers, the honey and pollen of which form the principal food of these 'Lorikeets', are as abundant there as leaves in summer are with us, and if supplied with its natural food in a warm and sunny clime, there is no doubt the Beautiful Parrakeet could be made to breed with little difficulty, for it is gentle and confiding, and soon becomes perfectly tame.

The Hon. and Rev. F. G. Dutton's account of the Psephoti.

PSEPHOTUS is the most delightful group of Parrakeets for aviary purposes. The four kinds I have kept are those which one generally sees: namely, *Hæmatonotus*, or the Redrump; *Hæmatogaster*, or the Blue Bonnet; *Multicolor*, or the Many-coloured; and *Pulcherrimus*, or the Paradise Parrakeet.

Of *Multicolor* I have not had much experience, having only kept a cock. This variety is the rarest, and, according to my experience, is quite as delicate, if not more so than *Pulcherrimus*: my bird was shy, too, like *Pulcherrimus*. He made no advances to tameness, and if I recollect rightly, ended by being found dead, without rhyme or reason, in his cage.

P. hæmatonotus, the Redrump, was the first I kept, and the first Paroquet I bred in a cage. I had them at Oxford in a cage some four feet long and three feet high. I provided them with an old candle-box, hitched on outside, and with a hole cut in the back. By this means I could always see how the nesting was going on. They laid two eggs, and took about seventeen days to hatch, at the end of which time they brought out one young one, which they successfully reared. Their nesting took place in the spring.

The cock bird was tame: the hen less so. He was very fond of poppy seed, and would come and pick it off my finger. So would the hen, but less readily. But the young one was as wild as a hawk, and so remained till one day it dashed out of the open door of the cage, and was lost to sight. Beginners who wish to try their hand at breeding Paroquets, can hardly do better than start with a pair of Redrumps.

P. pulcherrimus, the Paradise Paroquet, as dealers call it, is not only the most beautiful *Psephotus*, as its name says, but surely the most beautiful Paroquet that exists. The vivid emerald green and brilliant carmine of the cock, beautifully contrasted with the grey of the rest of the plumage, make him "a joy for ever." But "handsome is that handsome does", and I regret that I cannot give any of those I have kept a good character as a cage bird. They are very shy, and the cock is much given to driving about the hen. They do not appear to have been bred in captivity, but I do not think it impossible that they should do so. A pair I had were most anxious to burrow into the wall of a room in which they were. Had they done so, they would have got into a loft and escaped. So they were caged and sent to the Zoological Gardens on condition that they were to be turned into the Western aviary. I doubt if the condition was ever kept; for when I went to see how they were getting on, they were not there; no one seemed to know much about them, and after awhile I was told they were dead. It is not much use sending birds to the Zoological Gardens with a view to their being bred, for it must be recollected that they are zoological gardens and not a Jardin d'Acclimatation. If the Government would allow them more land, they could make the Gardens much more attractive, and of course more use; but as they are now restricted to the narrow and inconvenient plots they have, they cannot do more than they do: it is wonderful they do so much.

However, to return to my Paradise Paroquets, I regretted afterwards that a box covered with tin was not fastened on the other side of the wall into which they wanted to burrow: I think they might then have bred.

If I place the Paradise Paroquet at the head of Paroquets for beauty, I place the Blue Bonnet, *P. hæmatogaster*, at the head of Paroquets as a cage bird. It is the Merry-Andrew of birds. Who can describe its tricks in all their charm, amusement, and infinite variety? It is a bird full of resources, and never suffers from boredom. If it has nothing else to play with, it will play with its own tail. Really to enjoy them, they are better kept singly. If they are rather like *Pulcherrimus* in harrying their wives, they are the complete opposite of *Pulcherrimus* as to timidity. No birds could be bolder, and it needs a very short time to make them perfectly familiar with their master. I do not mean to say that they like being handled. Paroquets are not like Parrots and Cockatoos in this respect. The tamest of them endure handling rather than like it as a rule; and the Australian ones like it rather less than the Indian and American ones if anything. But they came forward to the edge of their cage, and are always ready for play. In fact if I were to have to choose which of all the species of the Parrot tribe should be the only representative of the family left on the earth, I should beg that it might be the Blue Bonnet. Mine did not make any advances to breeding.

I fed them all on millet, canary, and a little hemp. Oats would be good for them, or groats. They like plantains. They are very fond of bathing, and should always have a good-sized bath. I found Blue Bonnets rather subject to inflammation of the lungs. I never had any trouble with *Pulcherrimus* as to health.

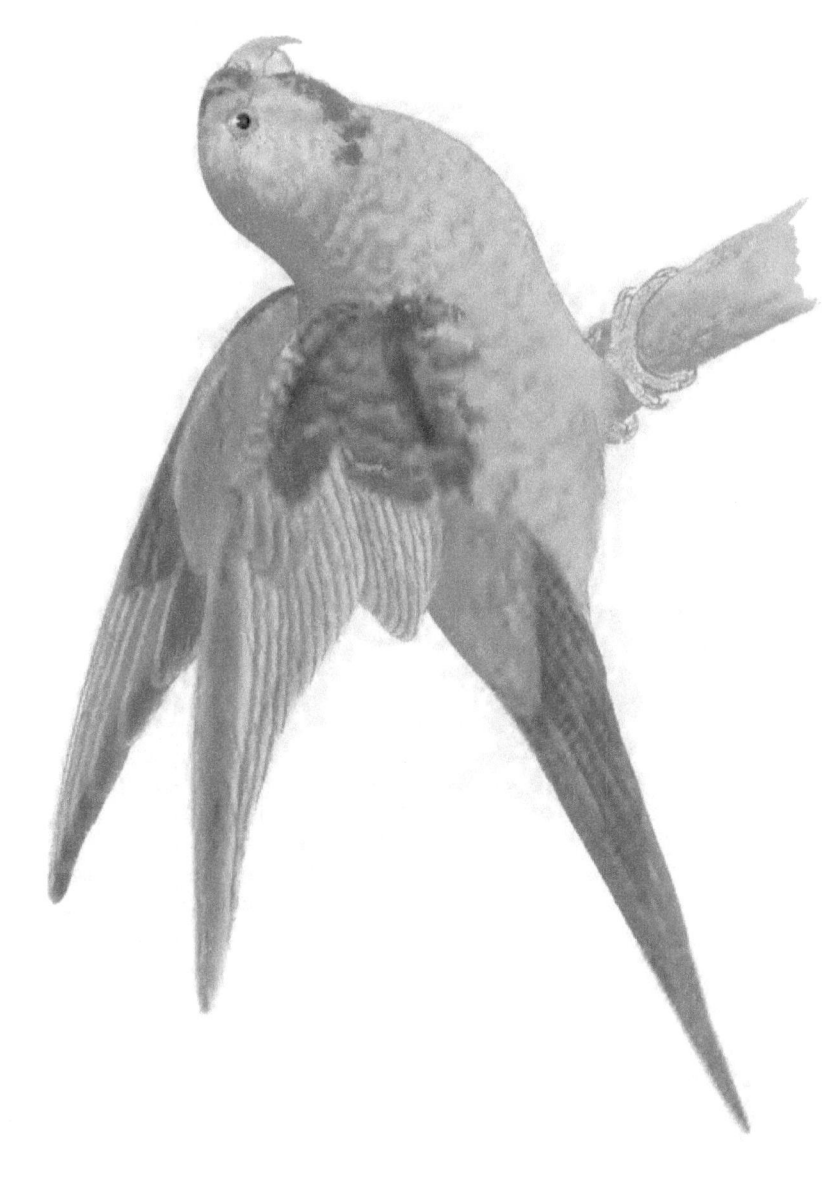

Swift Parrakeet, or Lorikeet.

Psittacus discolor, Russ.
Synonyms: *Psittacus humeralis*, Bchst.; *Psittacus Lathami*, Knl.;
Euphema discolor, Wagl.; *Lathamus discolor*, Gld.
German: *Der Schwalbenlori, oder der Lori mit rosenrothem Gesicht.*

NOTWITHSTANDING its English and German names, some authorities consider it more than doubtful whether this pretty little bird has any claim to be called a Lory: it eats seeds readily, and though not averse to sponge-cake, sugar, and mealworms, will live very well without these dainties, and is so hardy that it will thrive perfectly in a suitably constructed out-door aviary all the year round, which is scarcely to be wondered at, when one considers that its principal habitat, Tasmania, rejoices in a temperate climate, bearing much resemblance to that of the south of England, and that a considerable degree of cold prevails, during the winter, in both places.

Although common enough in their own country, the Swifts are not very frequently imported here; and when a few do occasionally arrive they always command a high price, which places them beyond the reach of ordinary amateurs.

Concerning this species Mr. Wiener writes as follows:—"Immediately after arrival these birds are delicate, and require careful feeding on millet and canary seed, and soaked bread or sponge-cake, to both of which a little honey may be added with advantage."

We have not found them delicate, and think that when landed in good condition they soon recover from the effects of the voyage, and, as we have said, become so hardy as to pass without inconvenience the winter out of doors.

The Swift is a pretty bird; the general colour of the plumage is bright green, a narrow band of scarlet marks the forehead, the top of the head is black, bordered with blue, the throat is deep red, the shoulders and upper part of the primaries blue, the under side of the wings red, the tail red with a blue tip, the breast and belly green with a yellow shade, flecked with red, the beak is reddish yellow, the legs and feet grey, and the eyes black.

The female bears a general resemblance to her mate, but is rather smaller.

The usual diet of these birds in captivity is canary and millet, but boiled oats may be advantageously added, and, during the breeding season, bread crumbs, coarse oatmeal, and a few insects of some kind, black-beetles or tipulæ rather than mealworms, but the latter, cautiously, in preference to none.

The Swifts make their nesting-places in the hollow boughs of the gum trees of their native land, and lay from four to seven white eggs, a little larger than those of the Budgerigar, and have usually two broods during the season, which extends from September to January. We have not heard of their having been bred in captivity as yet: in fact they are so seldom imported, that but few amateurs have had an opportunity of making the attempt.

In the Zoological Society's Gardens the Swifts are fed on boiled rice sweetened, and are not allowed any water; under such a course of treatment it is not surprising that there should frequently be a change of tenants in the cages allotted to them.

The price is about £3 a piece, or very nearly the weight of the bird in gold: it would consequently be well worth the while of any amateur to try and breed them, and this we intend doing as soon as practicable; for hardy as they are, when once fairly acclimatised, easy to feed, at least in comparison with many other species that are kept and bred without difficulty, none of the elements of success are wanting in their case.

While it is admitted on all sides by aviarists that immense pleasure and satisfaction is found to exist in the successful rearing of a brood of even the commonest birds of exotic origin that are kept in cages or aviaries, the enjoyment is enhanced ten-fold, nay a hundred, even a thousand-fold, when the fostering care of a painstaking amateur results in the appearance on the scene of a young family of rare and beautiful birds, which have never before been bred in England. What a wonderful amount too of pleasant gratification there is in penning a full, true, and particular account of that success for some bird-loving friend, or even for a gentle public enamoured of bird-life, and only too anxious to go and do likewise.

Well, everyone must make a beginning, and if the advice given in these pages is exactly and carefully followed, we flatter ourselves that much of the disappointment and failure that attended our early efforts in the pursuit of aviculture will be avoided, and success crown the attempt of even the tyro in the occupation, which is of such an engrossing nature that, we firmly believe, once it has been fairly taken up, it will never be entirely abandoned while life and health endure.

How we have digressed to be sure, and how far we have left our Swifts behind us! Well, they are such nice birds, and we were and are so anxious that amateurs should make a trial of breeding them in this changeable climate of ours, that the object of our digression will, we trust, obtain us pardon for its length.

The Swift Parrakeet differs considerably from its congeners in the shape of its wings, the primaries of which are narrow, and more than twice the length of the secondaries, consequently its flight is not only very powerful, but widely undulating in character; in fact so rapid is the progress of one of these birds through the air, that none but a most experienced shot could hope to bring it down.

The peculiar shape of the wings has caused more than one scientist to separate the Swift from the rest of the family, and constitute it a genus, of which it remains the only known species: but such minute distinctions are confusing and unnecessary, and have, very wisely, been discarded by many modern ornithologists, especially by Dr. Russ, who, recognising but one genus, distinguishes the various members of the Parrot family from one another, by specific names only; whether they be Cockatoos, Lories, Parrakeets or Parrots proper; their one generic appellation in the pages of his invaluable works is *Psittacus:* an arrangement that should at once commend itself to every thoughtful ornithologist, as there can then no longer be a doubt as to what family a bird with this prefix belongs.

In bird nomenclature, as in every other subject of popular study, simplicity and uniformity should, as far as possible, be the order of the day, and Dr. Russ has taken a right step in this direction, for which the thanks of all students of ornithology are due to him; and as his works become better known, and, consequently, properly appreciated, the horrible jargon, compounded of sonorous but too frequently inappropriate Greek and Latin words, will fall into well-merited oblivion, and birds be classed, as plants are, in "natural orders" rather than in genera, founded on trivial, or even imaginary distinctions.

The Swift is an example of the fact observed by many naturalists that while Parrots of the same species are found at great distances from each other when they are inhabitants of a continent, in islands each little sea-engirt morsel of land maintains one or more species peculiar to itself, and unknown even to other islets of the same group, however short the distance that separates them: this is peculiarly the case in many of the island-groups of Polynesia, and those that are contiguous to the great island-continent of Australia.

It having been lately denied that the Hobart Town Swift was a honey-eater, we may refer the reader to the late John Gould's account of this Parrakeet in his magnificent work *The Birds of Australia*, where he specially mentions having shot them in the vicinity of Hobart, as the capital of Tasmania is now called, and seen clear honey, to the extent of a dessert spoonful drop from their beaks when he held them up by their feet: and we have been assured of the same fact by other trustworthy informants, who had spent many years in the colony.

Since writing the above we have read an account of the Swift by M. Alfred Rousse, of Fontenay-le-Conte, who says (we translate): "This pretty Parrakeet is as hardy as possible, and deserves to be better known and more generally kept than it is. It bred with me in 1882, the first instance, I believe, of its nesting in confinement. I had had the birds in my possession since 1880. Incubation lasted twenty-one days, and in thirty more the young left the nest. This year again I have had a brood. The number of young to a nest (only one a year) varies from three to five. The sex of the young birds is at once apparent, as the red marks on the head of the males are already well developed."

Passerine or Blue-Winged Parrakeet.

Psittacus passerinus, Russ., Bchst., Linn., etc.
Synonyms: *Psittacula passerina*, Wgl., Bp., Fnsch., etc.;
Psittacula passerina, gregaria, et modesta, Lchtst., etc.   German:
Der Sperlingspapagei.

THIS nearly least of all the Parrot family, is truly deserving of the name of 'Love-bird', which is so generally bestowed upon other species, for it really is "inseparable", and must be bought and kept in pairs.

Known to amateurs from the time of Buffon and Bechstein, and probably from a very much earlier date, there are few birds more worthy of the notice of connoisseurs: scarcely the size of a plump cock Sparrow, the short tail makes it appear even less, and at the same time gives it a kind of unfinished look, that detracts somewhat from its personal appearance, otherwise so quaintly and quietly pretty.

"The little blue and green Parrakeet", says Bechstein, "is as social and affectionate as the preceding (the Red-faced Love-bird), but much more rare and dear."

"It is a native of Brazil", adds the old German author, "and cannot learn to speak."

The general colour of the plumage is deep green, the rump and the wings are sapphire blue, but scarcely to be noticed except when the bird is flying about, when the contrast of colours becomes very conspicuous.

The female is green all over, but with a whitish or greyish tinge about the head and neck, resembling very closely the female of the Madagascar Love-bird, no less than the female of the rarer Gregarious Parrakeet of the West Indies.

The species usually imported into this country comes from South

America, where it is sufficiently common, but we have frequently seen a bird that very closely resembled it in the woods of the basin of the Murray river in Australia, which, by the way, we do not recollect to have seen described by any author, Gould included, to whose works we have had access.

The Blue Wing has been very frequently bred on the continent, especially by Dr. Russ, who obtained it in the third and fourth generation. Our own birds have, as yet, only produced eggs, although one hen laid five, which were all fertile, and would have been hatched but for a thunderstorm that broke over the district when they were nearly ready to come out of the shell; whether the jar of the thunder killed the wee birdies, or frightened their mother from the nest, and so the eggs caught cold, who can tell? Yet she was not easily startled, and sat so closely that she injured her health by her devotion to the family that was not to be; and to such an extent that when we found the eggs were spoilt, and removed them, the poor little hen drooped and appeared so ill, that we took her out of the aviary with the double object in view of nursing her up, and at the same time preventing her from going to nest again until she was quite restored.

But alas! 'the best laid schemes gang aft agley': poor Blue Wing took the separation so terribly to heart, that after shrieking himself hoarse, and dashing wildly about the aviary, in vain attempts to find an avenue of escape, we turned her in again, when, so great was his joy, he fell off his perch in a fit, in which he very nearly died: but after a few minutes he seemed to rally himself by an effort, and managed to fly up to the perch upon which his wondering little wife was sitting, with something not very unlike tears in her usually bright black eyes, and pressed himself as closely as possible to her side: in this situation we left and found them after the lapse of some hours, when we discovered that he was almost incapable of supporting himself upon his feet, so we took the unfortunate little couple into the house, and—let us draw a veil over the close of the sad, sad tragedy.—How one's best intentions will miscarry now and then, and, where we meant nothing but good, carry woe and desolation to the very objects of our care.

Strange to say, not only does the male Blue Wing, fond as he is of his wife, not assist her in the weary task of incubation, but, as far as we could discover, even feed her while she is sitting on her eggs, which are usually four or five in number, roundish, and pure white: the period of incubation would seem to be about eighteen days; and there are said to two or three broods in the year.

These little birds are very frugal in their habits, preferring canary seed to any other, but eating freely of bread soaked in cold water, and squeezed nearly dry: they do not seem to care much for green food, but nibble a little fresh grass now and then.

They do not drink much, and we have not seen them bathe.

Their cry is harsh, and loud, but is not very frequently heard: the pairs converse with one another in a little subdued chatter, that sounds rather prettily, but they are usually silent and undemonstrative when under observation, which, as they are very quick, is not of frequent occurrence without their knowledge. We have watched one, through a loop-hole, for a quarter of an hour at a time, and have never seen him budge, his keen black eye fixed intently on ours all the while; and as soon as we looked away, he was off like a shot to the furthest corner of the aviary.

Although so timid at other times, the hen Blue Wing sits as devotedly on her eggs as any bird with which we are acquainted, even suffering her cocoa-nut husk to be taken down and carried to a distance without deserting her charge.

In their wild state these birds breed in the hollow branches of trees, but in the bird-room or aviary seem to prefer a medium-sized cocoa-nut husk for their nesting-place: they make no nest, properly so called, but content themselves with removing the superfluous fibre from the interior, and smoothing it down for their use.

They are better kept in a place by themselves, two or three pairs together, but are not to be trusted to the tender mercies of Budgerigars, Madagascars, or Red-faced Love-birds. A male, however, will mate with a female Madagascar (*Agapornis cana*), and even, a friend writes us, with a female Red-face (*Agapornis pullaria*), but whether the progeny of such unions, if progeny there were, would be capable of reproduction, as another acquaintance of ours is inclined to believe, we cannot say, though we doubt the fact; but if so, these three birds would be simply local varieties of the same, and not three distinct species as they are generally considered to be.

Will some of our readers make the experiment, and kindly acquaint us, in due time with the result?

There is another species of Love-bird that is often confounded with that under consideration: the resemblance between the two is so considerable that even Dr. Russ considers it an open question whether it is anything more than a variety of the ordinary Blue-winged Parrakeet; but we think it is. In the first place it is a decidedly larger bird, has a smaller beak, and the only blue about its person is on the under wing coverts, where it is not, of course, seen unless the bird is flying.

The female, too, differs considerably from the ordinary female Blue Wing, especially by her greenish yellow face, which is very perceptible when the birds are seen together. In a general way we object to the multiplication of species as leading to confusion; but where the differences are so marked as they are between the ordinary Blue Wing and the bird called *Der Sperlingspapagei ohne blaue Unterflügel* by Russ, the *Psittacus Santi Thomæ* of Kuhl, we have no alternative but to separate them, or abandon specific distinctions altogether.

Since writing the above we have observed that our widowed Blue Wing does not seem to be in the least affected by the loss of her poor husband, whose sad fate we have related on a preceding page, but on the contrary is getting over her moult nicely, and is looking strong and hearty, and we have no doubt will be quite ready to accept the attentions of a new husband when we present her with one in the Spring.

So much for the relative strength of attachment in the two sexes; the male died because he could not live alone for an hour, and the female, apparently, loses this constant mate unmoved, and bestows not a thought upon his memory. It seems almost a libel on the fair sex, but it is true nevertheless; the power of love is greater in the male than in the female, as a rule, throughout creation: there are exceptions no doubt, but these rather confirm than disprove our assertion.

"As far as my experience has gone", writes our friend Mrs. Cassirer in this connection, "I find the male birds, as a whole, far more sweet-tempered and devoted to their families than the hen birds."

Need we add another word?

GREY PARROT

GREY PARROT.

Psittacus erithacus, LINN., LATHM., BCHST., *etc.*
SYNONYMS: *Psittacus cum cauda rubra*, FNSCH.; *Psittacus Guineensis cinereus*, BESS.; *Psittacus ruber*, SEP.
GERMAN: *Der graue Papagei*, RUSS. FRENCH: *Le Perroquet gris, ou cendré, ou le Jaco*, BFFN.

IT is almost superfluous to describe so well known a bird, however not to do so would be *contra les règles*, so we shall proceed, briefly, to remark that the general colour of the plumage is a fine pearl grey, the feathers of the head, neck and belly are margined with whitish grey, and the tail, which is short, is of a bright vermilion colour; the beak is strong and black, the membrane at its base and the circle of the eyes have a powdered appearance, and on touching the bird an abundant chalky substance adheres to the hand; the feet are ash-coloured, and the iris yellowish; black in very young subjects.

Varieties of the Grey Parrot are not uncommon, and are generally supposed to come from the interior of the "black continent": two are mentioned by Brisson, one of which, according to that naturalist's account, has the wings marked with red, while the other has many red feathers mixed throughout the grey.

One of the latter sort was shown to Latham, and stated to him to have been brought from South America, an account of its origin which that writer very sensibly rejected; remarking, that if it had been brought to England from America, it had certainly been first imported from Africa, as no Grey Parrots were to be found in the dual continent, at least without some admixture of green in their feathers.

We have also seen a very dark, almost Black Parrot of this species, which we were assured was from Ceylon; but, like Latham, we rejected this account of its origin, and, while admitting that it may have been actually brought *from* that island of spices, we feel sure that it was not indigenous to it.

The male and female are exactly alike, and learn to talk with almost equal facility; the male, however, is usually the more fluent speaker of the two, but what few words the female learns she generally pronounces with great distinctness.

It has been said that a Parrot can only speak in one tone of voice, but this is not universally correct, for we once lived next door to a retired tradesman who was possessed of one of these birds, and of a white Pomeranian dog, which the bird would call "Carlo, Carlo", in such an exact imitation of his mistress's voice as to deceive the dog himself. The old gentleman was very fond of sitting and smoking in an arbour at the bottom of the garden, where his meditations were very frequently disturbed by his wife calling to him from the house, in order to consult him upon some domestic matter, on which occasions he used to call out, in the gruffest of gruff tones: "Well, what do you waant?" which the bird used to repeat so correctly as to leave the hearer in doubt as to whether it was the man or the Parrot that had spoken.

To Mrs. Cassirer we are indebted for the following account of a Grey Parrot, communicated to Dr. Brehm by a lady of high rank:—
"The bird of which I am about to relate some particulars, was presented to us by a man who had lived in the East Indies for a long time. The bird could already speak well, but only Dutch. In a short time, however, he learnt both German and French. These three languages he spoke as distinctly as a human being, and was so attentive that he often caught up expressions which had never been used before him; introducing them appropriately to the great astonishment of every one.

"He spoke single words and sentences in the Dutch language, and also introduced Dutch words with judgment between German ones, when he did not know, or had forgotten, the German equivalent. He asked questions and returned answers, made requests, and returned thanks; and used words correctly in relation to time, place, and persons.

"'Polly wants to kluk kluk (drink)': 'Polly wants something to eat.' If he did not receive what he wished for at once, he exclaimed: 'But Polly must and will have something to eat!' If he still received nothing, he flung everything about to show his anger!

"His morning greeting was 'Bon jour', the evening one 'Bon soir': he expressed a desire to rest, and took leave: 'Polly wants to go to sleep': as he was being carried away, he took leave, repeatedly saying: 'Bon soir, Bon soir.' Towards his mistress, who usually fed him, he displayed the greatest affection. When he received food from her, he kissed her hand with his beak, and said: 'I kiss the lady's hand.' He took great interest in everything his mistress did, and often, when he saw her busy about anything, would inquire, with extremely comical earnestness: 'Well, what is the lady doing there?' and when she was removed by death, and he no longer saw her, he also felt the loss and sorrowed. It was difficult to persuade him to take food, and to keep him alive. Often too he would re-waken the grief of the mourners, by asking them: 'But where is the lady then?'

"He whistled wonderfully, especially the tune *Ich dank dir durch dienen Sohn;* he also sang beautifully. 'Polly must sing a song', he would remind himself, and then begin:

> 'Perroquet mignon, dis-moi sans façon,
> Qu'a-t-on fait dans ma maison,
> Pendant mon absence?'

or the following couplet:

> 'Ohne Lieb und ohne Wein,
> Können wir doch leben.'

Occasionally he would alter this to,

> 'Ohne Lieb und ohne maison,
> Können wir doch leben.'

or he would substitute 'Ein Kuss—sans façon,' which amused him so much that he laughed loudly. 'Polly, what does Lotty say?' he would ask himself sometimes, and answer at once, as if some one else had asked him the question: 'O my lovely, lovely Polly, come and kiss me.' This was spoken with the correct accent of tenderness, as only Lotty could say it. His self-approbation he expressed by the words: 'Ah! ha! how lovely Polly is!' stroking his beak at the same time with his claw.

"He was, however, by no means beautiful, as he had the bad habit of plucking out his own feathers. Wine baths were ordered for him as a remedy, which were administered by means of a small syringe. These baths were extremely disagreeable to him; as soon as he noticed the preparations being made, he began to plead coaxingly: 'Don't make Polly wet—oh! poor Polly—don't—make—him—wet!' He was not fond of strangers; those who came to see him, and hear him speak, generally were only able to gratify their curiosity by hiding themselves from him. In their presence he remained as quiet as a mouse; but began to talk faster than ever as soon as they had taken leave, or concealed themselves, as if to indemnify himself for his self-imposed restraint. His affection, however, could be won, and he spoke willingly with such persons as were in the habit of visiting us, sometimes, indeed, making a clever joke at their expense.

"A stout Major, whom he knew very well, attempted one day to teach him tricks: 'Jump on the stick, Polly, on the stick!' commanded the warrior. Polly was extremely annoyed: then suddenly he burst out laughing loudly, and said: 'Major, jump on the stick, Major!'

"Another of his friends had not visited the house for some considerable time, this was spoken of, and it was expected that Roth, which was the name of the wished-for visitor, would come that day. 'Here comes Roth', suddenly exclaimed the Parrot, who had been looking out of the window, and had recognised the expected visitor at a distance.

"A son of the family, George, was expected home after a lengthened absence, and this was naturally talked about among the members of the household: George arrived late one evening, when Polly was already sleeping in the darkness of her covered cage. After the first greetings were over, George turned to the general favourite, and lifted the corner of the cage: 'Ah, George, art thou there? that is nice, very nice', said the bird.

"He had noticed that when his master went to the window, he often called to the steward, or to the bailiff, to come upstairs from the courtyard. When, after this, the Parrot saw his master go quickly towards the window, he called every time both the men by name, as he was unable to tell which of the two his master intended to summon.

"It is impossible for me to relate all that the bird did and said, he seemed almost a human being. Polly had a mournful ending. He was bestowed upon an aged relative of the family, who had become childish, and had taken a childish fancy to the bird. All wept as the wonderful creature was carried away: Polly alone shed no tears, but could not endure the parting from his beloved ones: a few days later he was dead."—Probably starved, poor thing, by his "childish" owner.

In the *Feathered World* for August 16th., 1883, a Mr. Diettrich relates the following anecdote of his Grey Parrot:—"It is very amusing to see Polly call the hens together, in imitation of my wife, and she then gave us no peace till she received a piece of bread, with which she took her seat on the window ledge, or on the paling, breaking it up and throwing down the crumbs. The running of the hens after the crumbs seemed to afford her the greatest amusement."

In the face of anecdotes like those related above, and others which are to follow, it would really appear as if these birds were gifted with a certain modicum of human intelligence; but a little reflection, and careful observation, will show that while the Parrot is certainly capable of attaching ideas to certain sounds, it is incapable of generalization; and that many of its most apposite answers, and remarks, are no more than mere coincidences.

A Parrot of this species belonging to a chemist in Bermondsey, where it is kept in the shop, calls out "Wanted", as soon as a customer comes in: and if the latter approaches the bird, and looks at it, it will put its head on one side and enquire, in quite a confidential tone of voice: "Well, who are you?" or "Well, what do you want?"

Another that belongs to a medical man of our acquaintance has learned, when a patient knocks, to say, "Open the door, and call the doctor", but occasionally it reverses this order, and shouts out, "Open the doctor, and call the door"; apparently quite unconscious of the mistake it has made: thus showing that although it may attach, and doubtless does, a certain meaning to the sentence it uses, the several words of which it is composed convey no ideas to its mind; and that this is really the case has been proved, in more than one instance, by actual experiment.

The majority of these birds that are sold in this country, are brought from the Gold Coast, but they appear to be pretty generally distributed throughout the western and central parts of Africa.

Buffon relates that in his time a pair of these birds bred for several years consecutively in Paris, and reared their young; this statement, however, has been questioned by some more recent writers, but is nevertheless probably quite correct, for a pair belonging to the late Mr. Charles Buxton, M.P., made a nest in a hollow branch, and "brought up two young Grey Parrots, which were afflicted with most awful tempers. The party of four fly about almost always together, and are a great ornament to the place" (Northrepps Hall).

The same gentleman continues: "A cat made her lodgings in one of the nest-boxes, and brought up her kittens in it, and two of the Grey Parrots, who had not been industrious enough to lay eggs and have a family of their own, were seized with the idea that these kittens were their children; they kept up a constant warfare with the old cat, and whenever she left the box one of them used to get in and sit with the kittens, and they were constantly in close attendance, even when the mother cat was at home."

"I had at one time", continues the same writer, "a flock of eleven Grey Parrots at my house in Surrey, but ten of them having got shot, the survivor associated himself with some Cockatoos, and for the last few years has invariably flown about in their company."

"The Grey Parrots have the sense to get into a house that was built for shelter for them, but none of the others can ever be persuaded to enter it: the gardener declares that the Grey Parrots foresee a storm, and often take refuge in their glass-house before it comes."

The foregoing extracts are from a paper read before the British Association in 1868, by the late Mr. C. Buxton, and nearly all the birds to which he there alludes, some fifty in number, fell a prey to "those vile guns"; one gamekeeper "bagged" no less than eleven, and, as Mr. Buxton good-naturedly put it, "naturally thought he had secured a wonderful prize."

"The Grey Parrot is a very good imitator of sounds and voices", wrote Mr. Sydney C. Buxton in *The Animal World* for 1878. "We had for many years an old retriever named 'Tory'—now, alas! dead of old age and merciful prussic acid. The Parrot could imitate our tone and call of 'Tory, Tory!' and when he happened to be in a merry mood (Parrots are fond of fun), and saw Tory half asleep, and comfortably curled up on the mat, he would call out 'Tory! To-ry!' The dog would rouse himself, anxious for a walk, look high and low, before and behind, and seeing no one, would begin to lay himself down again to rest, his temper slightly ruffled. Cries the Parrot, louder than before, 'Tory! To-ry!' Tory, now thoroughly roused, would glance about, and at last espying the Parrot, with a look of intense disgust and indignation, proceed to curl himself up again: the bird meanwhile chuckling to himself on the success of his practical joke."

Although the following anecdote from the same pen refers to a different species, we cannot refrain from quoting it:—"I spoke of the love of fun just now. We used to have a Grey Red-breasted Cockatoo, 'Minnehaha' by name, who would deliberately lay herself down on her back in the middle of the gravel-path, seized a pebble with one foot, fling it into the air, and catch it in her mouth if she could as it fell. All the while she would scream with pleasure and excitement, and evidently thought she was having the jolliest game possible."

The Grey Parrots in their native country feed on fruit and grain, principally maize, and many thousands of them die within a few weeks or months of their importation here. The causes of this mortality are various; fever contracted on ship-board, regret for the loss of their liberty, or their companions, sheer fright in some cases, and disgust at their surroundings in others, improper food in some instances, deprivation of water in some, and too much of it in others: but the chief cause of death is the inability of the young birds to feed themselves sufficiently to support life. On board-ship and at the dealers, when a number of these birds are caged together, the old ones feed the young ones, which require this attention for a good many months, for the Grey Parrot is a long-lived bird, and slowly reaches maturity. When removed from his companion, the poor young creature dies of slow starvation, and the disconsolate owner wonders, and buys another to meet, probably, with a similar fate.

The only way to preserve these young Parrots is to boil their corn until soft, chew a mouthful, and placing the beak of the bird in the mouth, let it feed itself there as it has been used to do from the mouths of its father and mother, or its kind companions in the dealer's shop.

There is a vile prejudice still existing in this country against giving water to Parrots; but we have already so fully descanted upon its absurdity, not to say wickedness, that we need merely here remark, that all animals drink, and can be kept without water only to their detriment and manifold discomfort: but the water must be fresh and clean, that is a *sine qua non:* foul water means diarrhœa, inflammation of the bowels, fever, and death.

The Grey Parrot, as we have remarked, grows slowly, and attains to a green old age: some specimens are reported to have lived for sixty, eighty, and even one hundred years, but the truth of this statement we are unable to vouch.

Apropos of the bird under consideration, a writer in a recent number of the *Daily Telegraph*, under the heading *On the Congo with Stanley*, says: "Flocks of Grey Parrots flow across the sky, alternately screeching and whistling melodiously. I have seen it stated erroneously that the Grey Parrot never whistles in a wild state. On the contrary, it does so very sweetly, and with a great variety of note."

Well, one certainly lives and learns: it is comprehensible nevertheless that the sibilant utterances of *Erithacus* in a state of freedom may be devoid of the concentrated bitterness that usually marks his attempts at vocalization in captivity, when his temper has been spoiled, and his digestion ruined, by alternate teasing and stuffing with inappropriate tit bits; or the writer of the above quotation may, by the sight of the wild birds, have been pleasantly reminded of some familiar "Polly" of his acquaintance, and the associate ideas connected therewith, have lent a melody to the Parrot's notes they might not otherwise have possessed: we never saw any wild Parrots (we do not include Parrakeets) that did anything else but scream horribly; but then, of course, it does not follow that others may not have been more fortunate, and we certainly have not been on the Congo, wandered on the shores of Stanley Pool, or gazed on the luxuriant vegetation that adorns the islands dotted on the surface of its limped waves: "palms beautiful and symmetrical, with hanging clusters of bright orange-coloured fruit, masses of yellow flowers, lilac-coloured *papilionaceæ*, and mauve convolvuluses, beautiful scarlet seed-vessels of a certain bean that form blazing clusters of gorgeous effect amid the tender green foliage"; nor have we been happy enough to behold the "immense numbers of Grey Parrots, small flocks of them going together that flutter and play about the tops of the tall trees, whistling and screaming joyously all the time", or been privileged to see "the many snags that rear their withered branches over the rushing stream, where numerous little birds have for safety's sake hung their pendant nests of grass, so that there is a constant twittering and fluttering of pretty and brilliant forms round the gnarled old trunks and whitened twigs"—a lovely scene surely, and a description that inspires the reader with a desire to start off forthwith and feast his eyes upon its unparalleled beauty.

The following interesting particulars from the pen of Mr. J. G. Keulemans will be read with interest:—"Of all the foreign cage-birds that decorate and enliven our dwellings, few are more common or better known than the Grey Parrot. Large numbers are being continually brought to Europe from their native wilds, and at some places—Lisbon, for instance—they may be seen in large numbers at the bird-shops, but nowhere is the Grey Parrot more frequently found as a cage-bird than in London.

"The range of the Grey Parrot is limited to the Western Coast of Africa, and extends for some distance into the interior. It is common on the Gold Coast and adjacent islands; but is curiously distributed among these latter. On Prince's Island we find these birds in great abundance, while on the neighbouring island of St. Thomas not a Grey Parrot is to be seen—a fact to be accounted for by the large numbers of the Kite (*Milvus parasiticus*) inhabiting the latter island.

"Although a familiar cage-bird very little is known about its habits when in the wild state. It is therefore with much satisfaction that I find myself able, from personal observation, to communicate many new and interesting particulars concerning it, which I hope may prove acceptable to my readers, and at the same time enable them to form some idea of the *vie privée* of this favourite.

"At Prince's Island, which may not inappropriately be termed the Paradise of the Grey Parrots, I resided for more than a year, and during that time I daily carefully observed their habits and mode of life in the natural state. Nowhere on the continent of Africa are these birds so plentiful, nowhere so free and undisturbed. On Prince's Island they are supreme among the birds; they stand in no dread of the other feathered inhabitants, but are feared and respected by them.

From their own immediate domain the Parrots drive away all other birds, both great and small,—if necessary combining for that purpose.

"The only enemy they meet with are the Kites (*Milvus parasiticus*) of the neighbouring island of St. Thomas; it sometimes happens that a Kite does, either by design or accident, find its way to Prince's Island, but no sooner is the intruder observed than the alarm is raised, the Parrots hasten up from all parts, and in a very short time the luckless bird is either killed or driven away.

"During the day, when flying about in flocks, the Parrots never settle on a tree, unless satisfied that it is a safe resting-place. They are very suspicious, and always on the alert, taking notice of everything that occurs in their vicinity. They are more prudent and sharp than the native, quicker than the monkey; they require no tools to crack the hard nuts, and are consequently the most independent of the living creatures on that island.

"On Prince's Island there is a very lofty mountain, reaching some 1200 feet above the level of the sea, and called by the natives 'Pico de Papagaio', or Peak of the Parrot. On the slope of this mountain, and extending far up its side, is a magnificent forest. The trees are of great size and height, and their trunks and branches give support to the lianos and other climbing plants, which hang about them in rich luxuriant folds. The density of the forest is so great that it is only with the utmost difficulty and toil the explorer can force a passage through it, while to the Parrots, who come up there every night, it presents no obstacle, but gives them, under the shelter of its thick foliage, a secure and pleasant resting-place.

"As sunset draws on, the Parrots may be seen in parties winging their way for the mountain from all sides of the island, and on reaching it take their places on the trees. Approaching troops acquaint their fellows of their coming by a loud whistling. Those of them who have found an approved resting-place warble and whistle as long as daylight continues, but as darkness closes in the noise gradually subsides, and all becomes hushed. Occasionally, however, a few sounds may be heard at intervals after dark, which most probably proceed from some belated bird seeking a place or a quarrel: sometimes in the dead of the night the whole colony is thrown into an uproar, occasioned, I believe, by the visit of bats or of some predacious animal.

"There was one flock in particular, consisting of about forty individuals, that attracted my special attention; every evening at nearly the same time, namely, half-past five, they would pass over my house on their way to the mountain. I used to follow them with the eye, and always found that they settled on the same tree.

"My house and plantation were situated at an elevation of some 1200 feet above the sea-level and opposite to the Pico de Papagaio, a valley being between. From here I had an extensive view across to the Pico; and observing a trail in the direction of the tree which this particular flock of Parrots had taken for their resting-place, I determined to pay them a visit. As the distance did not seem to be very great, I thought that by following the path visible from my house I should have little difficulty in reaching the wished-for spot without either guide or assistance.

"Accordingly, on the 16th. of January, 1865, I set out unaccompanied, at daybreak, for the place in question, and soon reached the edge of the forest, but before I had penetrated very far it became evident that my plan was not so easy of execution as I had imagined, for that which from my window looked like a path, turned out to be merely a rough track or trail overrun with rank vegetation, which only served me for a short distance, and then became obliterated; this loss of track brought me to a standstill. I was alone and had no knowledge of the way, but being unwilling to return, having got thus far, I looked about in hopes of discovering in this dense tangle some place through which I might force a passage, but in vain! on all sides the lianos and other climbing plants grew so thickly, and presented such a high impenetrable barrier, that I felt to attempt to proceed further without a guide would be unsafe, and at once decided to retrace my steps and obtain the assistance of a native. From surrounding appearances I was convinced that many Parrots were breeding there already, and the idea of giving up a trip that promised such interesting results was not to be thought of. I hastened back and soon reached my house again. Having secured the aid of a native, who assured me that he was able to make his way to the place and find his way back, I set out again the same morning with the hope that this time my wish to reach the abode of the Parrots would be realized.

"At nine o'clock we started, and after a very fatiguing journey, during which we had in many places literally to cut our way through the dense undergrowth, reached one of the places frequented by the Grey Parrots. My gratification at this moment was extreme. What should be done? Shoot as many of the old birds as presented themselves, or seek out their nests and take the young ones home? We decided to take the young birds. From the noise we knew that many Parrots were in our vicinity: on all sides their joyful whistling resounded, and the falling of pits, or husks, and the stones of fruits, sufficiently proved that many were concealed in the trees about us. Up to this time we had not seen any; they kept themselves so well hidden among the leaves that we could not get sight of them. Thus they remained for some little time, when at last one came down, and I could not restrain the temptation to shoot it. At the discharge of the gun a multitude suddenly burst from their concealment and dispersed with deafening screams. Picking up my prize we walked cautiously forwards until we observed at a distance in front of us another troop—or it might have been the one we had disturbed—on the highest branches of an exceedingly large tree. In half an hour we reached the spot, where we found numbers of large trees with foilage so thick that we could not see any Parrots. While waiting and eagerly watching we again heard the falling of empty husks, and at the same time observed a movement among the leaves of a palm tree. On closer inspection we could see our birds, and as we had no desire to cause another Parrot revolution we left them to eat their palm nuts in peace. It being now about five o'clock in the afternoon, and the sun setting at six, we had only an hour left at our disposal to seek after the nests. My companion being more likely, from his experience, to spy out the nests quicker than myself, I set him to inspect the trees within a short circuit, while I remained at that spot, and arranged that he should inform me of the discovery of a nest by imitating the call of the Wood Kingfisher (*Halcyon dryas*). Soon after the sudden signal of my black man informed me that he had been successful in finding nests: repeating the call alternately, I at last got up to the place where he was, and the position of the nest was pointed out to me. A hole in the trunk of one of the trees was, according to my man, the home of a pair of Parrots. Knowing from experience how sharp-eyed these fellows are, I was quite sure that something Parrot-like would be found inside the hole. Darkness was now coming on fast, and it being too late to do anything that night, we marked the tree by fastening some palm-leaves on the trunk, and left the breeding pair inside the hole undisturbed till next morning. While thus occupied, troops of Parrots approached from different sides and settled among the trees. As for ourselves we found a capital shelter under a clump of bushes, where we concealed ourselves, and from there observed unperceived the doings of the Parrots around us. Some were climbing and hanging on the branches, others flying and scampering through the foliage; we saw them perching close to each other, and afterwards five or six settled just above our shelter. Numbers came from all sides, and the chattering which we had previously heard at the distance by was this time close to us. There was a noise of whistling, screaming, quarrelling, and the breaking of dead branches. We saw them pass before us and settle on the trees: at this time we must have been surrounded by hundreds of Parrots. Being now almost dark, and having to pass the night in the open air, it was time to take steps to make our sojourn in the forest as comfortable as possible. A fire being the first necessity, we left our shelter in order to gather some of the dead palm-leaves that lay about; as we emerged from our coverts the Parrots perceived us, and in a moment the whole place was ringing with their deafening screams. The fire was soon made, and burning up quickly, it cast a cheerful light and warmth around the spot, rendering our bivouac more agreeable; and the Parrots, attracted by the unusual sight, kept flying over and around the place thus illuminated. An hour afterwards, quiet been restored, we proceeded to get our supper, consisting of roasted bananas; this being finished, we dried some moss to serve for a bed and retired; but the night was so cold, and the mosquitos kept stinging my face so pertinaciously, that it was impossible to sleep; so I got up and roasted some more bananas, smoked a pipe, and then felt quite ready to go in pursuit of the birds. My companion was sleeping soundly, apparently undisturbed by those causes that deprived me of my sleep. As it wanted some three hours of daylight I occupied myself in preparing limed sticks and making snares. On the appearance of the first streak of dawn we proceeded to the tree where a nest was suspected to be; my black man, being a capital climber, went up to the hole, and looking in found two young Parrots, which he carefully tied up in a pocket-handkerchief and lowered down to me: the little things seemed to be about a fortnight old, and readily took some banana I offered them. The old birds were absent, probably seeking food; at least we did not see them. The two young ones we fastened with string to the trunk of the tree, and placed some limed sticks round about them. In this way we caught one of the parent birds, the securing of which was a matter of some difficulty, owing to the resistance it made. We put the freshly caught old bird in a linen bag, and fastened it beside the young ones. In a few minutes our captive began to turn round and round, at the same time screaming vociferously: this quickly attracted the attention of the Parrots in the neighbourhood, which came near, gazing with astonishment at the bag which contained their mysteriously hidden comrade. One more bold than the rest flew down and settled on a lime stick, but its struggles were so violent that it succeeded in getting away. I then took my gun and shot two individuals, the others immediately taking flight.

"On the same morning we discovered two more nests: one of them contained three very young birds; the other had only one egg. My man had previously laid some snares, but no birds had been caught. We then searched about among the trees in this part of the Pico de

Papagaio, and discovered several holes, many of which probably contained nests, but we were not able to climb up to see, the trunks being so smooth and thick.

"We set out on our return for my plantation at noon, and arrived there safely at 4 p.m. Although my excursion had furnished me with the small result of six living Parrots, three dead ones, and one egg, yet the exertion this had entailed brought on a severe fever, which developed itself as soon as I got home and confined me to my house for several days.

"While on the Pico I availed myself of such a favourable opportunity, and took particular notice of the habits of this bird. The first thing I observed was that it is always found in flocks, which flocks go about over the island during the day, returning to their own appointed place on the mountain in the evening to roost. Their food consists of fruits, such as the palm nut, the avocat (*Laurus persea*), the banana (*Musa paradisea*), goyave, mango, and many other fruits of a smaller kind, but they always give the preference to palm nuts.

"They drink but little, and as no water is found on the Pico they must obtain what they require during the day on the lowland. They make no nest, but deposit their eggs (which are from two to four in number) on the bottom of the hole. The eggs are in size, shape, and colour similar to those of the Wood Pigeon (*Columba palumbus*): when unblown they are of a pinkish hue, which may be owing to the thinness of the shell. Both birds take it by turns to sit, and while one is sitting the other often comes and feeds it out of its crop. The young ones are fed in the same way. In time of danger the old birds defend their progeny vigorously, and should the enemy prove too strong to be successfully resisted by one pair, other Parrots come up to their assistance, and joining forces either kill or put the aggressor to flight.

"The Grey Parrot delights to dwell in companies: many nests are found within a few feet of each other, and often in one tree two or more holes may be seen occupied by hatching pairs.

"The young birds are covered with a long and fluffy down, which afterwards, when moulting, falls off. Their first plumage is darker, and the iris dark grey, instead of pale yellow. They leave the nest when about four weeks old, but may be seen looking outside the hole some time before they are able to fly. They grow quickly, and the feathers get gradually paler; when two months old the first moulting begins, which lasts about five weeks, after which the plumage is similar to that of the old birds, although the edges of the feathers are not so pale and the cheeks and forehead not so white as in old individuals.

The iris changes gradually and slowly; the eyes are dark for more than seven months. The feathers when wet appear of a dark bluish grey, with a purple gloss.

"As to the method of treatment, I have always found hemp seed to be a very good kind of food, and one that is relished by the bird; boiled rice is also very suitable, and a lump of sugar is a source of great amusement. They will eat canary and other small-grained seeds, but these I think are objectionable, as they keep the bird too much occupied with eating, so that they lose a good deal of their capacity for imitating words. The best food is Indian corn boiled or ground and mixed with hemp seed, and bread softened in sugar water. Almonds are not easily digested, and bitter almonds make them ill. Parsley is poison for a Parrot. Fresh fruit, such as figs, pears, cherries, etc., always give great satisfaction to the bird, but they must not have too much of it."

The Hon. and Rev. F. G. Dutton's account of the Grey Parrot.

Of all Parrots there is none so well known as the Grey. If a man has ever seen a Parrot, it is probably a Grey that he has seen. If a person wants a Parrot, he wants a bird that will talk, and the Grey Parrot is, on the whole, the surest of talkers. Other Parrots may become more wonderful talkers, but they probably will require careful teaching, the Grey Parrot can be trusted to pick up words without much trouble.

But in spite of this, a great deal of misapprehension exists as to the talking powers of the Grey Parrot. One hears stories of Grey Parrots that had been in the possession of different people, and would suddenly pass from a childish treble to a gruff man's voice. Such birds may have existed, but I can only say that though I have possessed several Grey Parrots, and known more, I never yet came across one that had this power. The Macaws unquestionably possess it; Amazons have an unrivalled power of imitating the general tenour of a conversation without pronouncing one distinct word, but Grey Parrots, according to my experience, though they learn to pronounce their words distinctly, always do so in their own rather throaty voice: nor are they by any means all gifted with the same power of speech.

I have had a Grey Parrot which, though very tame, could never learn a word; and I had another which only required a week or two to learn long sentences, and began to reproduce them after a day or two. Books tell one that the two sexes talk equally well. I doubt this about any species of Parrot, or bird. Just as in song-birds, the male is the songster, so in talking-birds, I believe the male to be the talker. The female may learn to speak, but not nearly so well, as a rule. That there are exceptions is as probable as that there are hen Canaries with a considerable power of song.

But how to distinguish the sex of the Grey Parrot, I know not. Some people say that the hen has the flatter and broader skull. I cannot say: but I can say this, which gives colour to the assertion, that all the good talkers I have had, had small neat heads, with the skull nicely arched and a small beak; while the flatter headed birds with large beaks have been the bad talkers.

Certainly if I were going to buy a young one, I should take care to pick out one with a small arched skull, and with a small beak.

This brings me to the subject of the purchase of a young Parrot. Many of those imported, perhaps it would not be too much to say most of those imported, die soon after their arrival, or soon after their purchase. Many of them probably have already the seeds of disease in them, and the journey from the dealer to the purchaser's house takes away what little chance they might have had of recovering. The chief symptom is an unquenchable thirst and diarrhœa. To give the bird *cold* water will add to the mischief. I should try *milk*—skim milk—with a little brandy in it—a teaspoonful to four tablespoonsful. If it could not digest that, then I should try warm water and brandy. By this means, if the bird will do nothing but drink, you are keeping up its strength all the time.

I should occasionally vary this diet with warm chicken broth, till I could get the bird to eat boiled maize or rice. It must be kept in an even warmth, about 70°; for Grey Parrots, heat alone is almost a cure for many of their illnesses.

In old days, the books always used to give bread and milk as the proper food for Parrots. I believe very few adult Parrots will be found to eat it, and it is not very good for them, if they will. But young Parrots require it, and it may be continued till they showed a distaste for it, when it can be changed for bread and water, or for plain water. As to the seed for them, hemp, canary, and millet, are all good. The three best and plumpest Grey Parrots I have ever seen were fed, one on nothing but hemp, and the other two on nothing but maize unboiled.

Sometimes one meets with Grey Parrots with the red feathers appearing amongst the grey. These used to be thought accidental examples, and were said by their owners to be unrivalled talkers: a recent traveller, however, reports that there is a district in Africa where the variegated bird replaces the ordinary type of grey. If that be so, it is to be hoped that some may be brought to England.

In conclusion, I cannot too strongly impress upon the reader the necessity of giving the birds plenty of employment,—sticks to bite, or reels to play with, if they are to be kept from feather plucking, which generally arises from want of something to do.

P.S.—Since writing the above, I have seen a Parrot which, according to its owner's testimony, can perfectly imitate two distinct voices. I have not heard it do so myself, but I have not the least reason to doubt that it does.

As we have already stated, we have known a Parrot that could imitate exactly the voices of both its master and its mistress so accurately as to deceive every one that heard it.

SENEGAL PARROT.

Psittacus senegalus, LINN., LTHM., KHL., *etc.*
SYNONYMS: *Psittacus senegalensis*, BR.; *Psittacus Versteri*, GFF.;
Psittacula senegalensis, BRSS.; *Pionus senegalus*, WOL.;
Poicephalus senegalensis, SWNS.;
Poiocephalus senegalus et senegalensis, HRTL.; *Phoeocephalus senegalensis*,
BR.; *Pionias senegalus*, FNSCH.
GERMAN: *Der Mohrenkopf-Papagei*, RUSS.

IN different parts of the world, Africa, America, and Asia, are to be found birds which have received their names from the fact of their being possessed of black heads, but the species now before us, "known in Europe from the most remote period", as Dr. Russ says, "is a handsome bird", and by far the most desirable of them all as a pet.

"These birds", to quote further from Dr. Russ's description, "must be reckoned among those which have been known in Europe from the earliest times; at present they are regular guests in all the Zoological Gardens, and tolerably frequent at the dealers. The first pair which I kept in my bird-room were indescribably wild and obstinate; at every approach or even the least movement on the part of the observer they uttered disagreeable piercing shrieks, and every attempt to tame them proved fruitless. However, they took possession of a nesting-box, at first probably as a place of refuge in their timidity and wildness, and commenced nesting. Three very round and disproportionately small eggs were laid, but were soon eaten up by the male; and when the hen laid again and commenced sitting, the monster bit her dead and gnawed her skull.

"This Black-headed Parrot, however, was so handsome that I could not make up my mind to get rid of him; and when, after some months of solitude and at least partial taming, I gave him a second hen, he proved much more amiable: they did not, however, breed.

"Sometimes a young Black-headed Parrot will become very tame, and, according to a communication from Herr Von Schlechtendal, also very amiable; while others assert that he learns to speak a little. From fear and nervousness he utters a curious grating sound, but when greatly terrified a shrill, whistling scream. During the breeding season he executes a strange love-dance."—*Handbook*, page 246-7.

To a friend, Mrs. Cassirer, of Paris, we are indebted for the following translation from the same author's *Die Sprechenden Papageien*, of a further account of the bird under consideration:—

"The Orange-bellied Long-winged Parrot, (*Psittacus senegalus*, L., *Mohrenkopf-Papagei*, *Perroquet de Sénégal*, *Perroquet à tête noire*, *Senegal Langvleugel-Papegaai*,) the pretty Mohrenkopf, as it is almost exclusively termed, belongs to the commonest birds of commerce, and reaches us regularly every year in considerable numbers. It must also be included among those birds which have long been known to us, for it is mentioned in 1445 by Aloysius Cada Mosto, and described by Brisson in 1760.

"The adult male bird is either brownish or blackish grey on the head, cheeks, and upper part of the throat; the back, rump, and upper tail coverts are glossy grass green, the pinions olive green brown; the wing coverts green with brown centres; the shoulders and small wing coverts on the under side are yellow; the tail, and all the rest of the upper side are bright grass green, which is also the colour of the throat and upper breast; the remainder of the under side is yellow; the breast and belly are orange, or the colour of red-lead; the under tail coverts are yellow; the beak is dark horn grey, merging into blackish brown; the cere is blackish, the iris sulphur yellow, to dark brown; a blackish ring of naked skin surrounds the eye, the feet are blackish brown, and the nails black.

"The female has the head a light brownish grey, her underside is a uniform yellow, without any orange red; her under tail coverts are yellowish green; and in other respects she resembles the male, but is smaller; in size she scarcely equals a Daw. Length 26—28 c., wings 14.5—15.7 c., tail 5.7 to 6.7 c.

"Habitat West Africa and Senegambia, but probably extends deep into Central Africa. Very little information has been received hitherto as to its mode of life while at liberty. In families of about six in number they frequent the gigantic monkey-bread trees, and betray themselves at every approach by piercing screams. Though awkward at rising up, and settling down, they fly swift as arrows. The mode of nidification has not yet been observed. After the nesting season they wander about, and at times do considerable damage to the banana, rice, maize, etc., crops.

"The Black-headed Parrot is chiefly important as an ornamental bird for zoological gardens, or particular amateurs. Although the earlier authors unite in declaring that these birds have no power of speech, the contrary has been proved in many cases; of late, accounts have been given of specimens which had learned to speak. Such descriptions have been recorded by Herr Von Schlechtendal in Merseburg, and Herr Fielder in Agram, and also by Herr A. E. Blaauw.

"Old birds of this kind are extremely wild and unmanageable; amid piercing screams he flings himself head downwards at every approach, squeezes himself into a corner, and utters a curious grating noise; young birds, however, are soon tamed and are very docile. One of these birds was able to open every door, and was fond of playing, and extremely droll, also very good-tempered and confiding, permitting his head to be scratched, allowing himself to be taken out of the cage and caressed, but he only learned to speak a few words, though he imitated the notes of other birds.

"Herr Blaauw relates that his Black-headed Parrot spoke French very prettily, very distinctly and softly. 'It has a strange effect, when he mixes the different words and sentences with his natural notes, and thus screams with articulate sounds.'

"Immediately after importation, even this otherwise strong and hardy Parrot has shown itself to be very delicate, at least of late: he is apt to fall ill on every change of diet, especially, it seems, if too lavishly supplied with hemp. At first therefore it should receive only canary seed and oats, and later, by degrees, hemp and sunflower seeds. A small supply of sound, sweet fruit is also necessary.

"A freshly-imported Black-headed Parrot can be bought at from fifteen to twenty shillings; tame ones from twenty to thirty shillings. I cannot name a price for a speaker, since such must still be considered as rarities, and scarcely obtainable in the trade."

As we write these birds are offered very cheaply in the English market, namely at prices ranging from five to six shillings a piece, and are chiefly imported *via* Liverpool, where they sometimes arrive in immense numbers; but as they are mostly quite young, very many of them die soon after arrival, especially when purchased singly.

Owing to the fact that the name of Senegal Parrot, *Perroquet de Sénégal*, has been given to several birds, a certain amount of confusion exists as to the identity of each; but the Mohrenkopf, we consider, more justly entitled to the designation than the other species which are described by several authors under the name which, following the example of Dr. Russ, we have preferred to restrict to that under consideration.

The name of Senegal Parrot, or Parrakeet, is very often given to the black-beaked African variety of the Ring-necked Parrakeet, *Psittacus torquatus, var. docilis*, but erroneously so, in our opinion, as it is much more appropriately bestowed upon the subject of the present notice, which really belongs to that part of the "dark" continent, whence it is occasionally imported in considerable numbers *via* Hâvre and Bordeaux: there is, at present, only one representative of this species in the Gardens of the London Zoological Society, where, on the whole, the Parrot race is very inadequately represented, and the worst accommodated of all the animals belonging to their collection: but now that attention has been called to the subject, let us hope that the Society will exert itself to cause the speedy removal of an approbrium that does much to mar the reputation they have justly obtained in other branches of Natural History: it is true, as our colleague has remarked, that the Zoological Society is not *une Société d'Acclimatation;* but nevertheless they should do something to enhance the comfort of the poor Parrots committed to their charge: they have handsomely provided for the obscene apes and monkeys of their collection, and more recently for the snakes and crocodiles, and it is only fair that the Parrots should have their turn, and we sincerely hope that, for the poor creatures' own sakes, it may come ere long.

Hyacinthine Macaw

HYACINTHINE MACAW.

Psittacus hyacinthinus, Russ.
SYNONYMS: *Ara hyacinthina*, GR., SCULGL.; *Psittacus Augustus*, SHW.;
Psittacara cobaltina, BRJ.; *Macrocercus Augustus*, STPH.;
Macrocercus hyacinthinus, LSS.; *Anodorhyncus Maximiliani*, SPX.;
Ara hyacinthinus, FNSCH.; *Sittace hyacinthinus*, WGL.
GERMAN: *Der hyazinthblaue Arara.*

THIS is a very rare bird, possessed by a few Zoological Gardens only; its general colour is deep blue, and it is, as Dr. Russ remarks, distinguished by a particularly colossal beak (*mit besonders kolossalem Schnabel*).

A very fine specimen has survived for a considerable time in the Gardens of the London Zoological Society, where it has learned to repeat a few words, and is especially partial to the youthful visitors, who, with no lavish hand, share their buns and cakes with it, as it screams and swings just above their heads on the perch to which it is chained under the trees, by the Parrot-house, facing the Regent's Canal.

The usual diet is maize, hemp, monkey-nuts, to which may be added biscuits, nuts of all kinds, apples and fruit; it is one of the few Parrots in the "Zoo" that is permitted to drink, and certainly appears to thrive on the regimen provided for it.

We do not admire any of the Macaws, and would not be tempted to keep one of them for a good deal; still we cannot quite agree with Mr. Wiener that "their huge size, brilliant feathers, and loud screams are a very good advertisement for a travelling menagerie, to whom amateurs had better abandon these birds, unless some one would care to construct a wrought-iron in-door aviary (I doubt whether bricks and mortar would be proof against their beaks) to make an attempt at breeding."

Dr. Russ quotes the price of one of these birds at from six hundred marks to nine hundred marks—the mark being about equivalent to an English shilling.

This bird is stated, on the authority of Azara, to depart from the general habits of the family in selecting a nesting-place, and instead of rearing its young brood in a hollow tree, to scrape out for itself a burrow in the bank of some stream; also to lay but two eggs to a sitting, and to rear two broods in the season.

It would be curious to find out the reason, or reasons that have compelled the departure of this bird from the habits general to the greater number of its congeners, but it is hopeless to make the attempt, unless some one should acquire a knowledge of the creature's language, and obtain a personal explanation from the bird itself; it cannot be from lack of hollow trees in which to breed, for the Hyacinthine Macaw inhabits the same regions as many of the tree-nesting Macaws, the Amazon Parrots, and the Toucans; and it can scarcely be that the banks of a stream, in a country subject to inundations, afford a securer dwelling-place than the hollow trees in which so many of its relations live.

There is no rule without an exception, it is said, and it probably is in order to prove the rule that Parrots build in hollow trees, that the Hyacinthine Macaw, and a few others, have selected for themselves a dwelling place of a totally different character.

We have no knowledge of these birds having bred in captivity, but from indications we have observed in the specimen living in the Gardens of the London Zoological Society in the Regent's Park, we should say there would be no difficulty in inducing them to breed, were they but provided with suitable accommodation, in a dwelling-place of sufficient extent to contain a stream with a bank, and a hollow tree or two, when it would be extremely interesting to observe on which of the two situations they would fix their choice.

Will some one, at the "Zoo", or out of it—preferably out of it—make the attempt, and let us know the result. We would do it, but unfortunately have not the necessary accommodation: but there are plenty of rich amateurs to whom the expenditure of £50 or so in the gratification of their peculiar hobby is no object at all, let some of these try what they can do, and determine, as far as can be practically done, whether it is by choice, or from necessity, that the Hyacinthine Macaw makes a burrow in a bank, instead of in a tree, for the purpose of rearing its callow brood; for it is only by thus experimenting that this and other kindred and equally interesting questions can be solved.

Though noisy, the Macaws, and the Hyacinthine species in particular, are fairly intelligent birds, and may be taught to speak, not only single words, but even short sentences: the specimen in our Zoological Gardens, for instance, always shouts out, when he sees us approaching him, "Come along, come along", and occasionally, "Hollo there! give us a piece", or words to that effect; so that, if we had accommodation for them, we should feel inclined to try some of them for breeding; but surrounded as we are by neighbours, most of whom have no sympathy with our ornithological pursuits, we feel that to attempt to keep any of these fine birds is simply impracticable, for they are so terribly noisy, that a summons or two to appear before the County Court Judge as a nuisance would be certain to greet us before long, and we have no desire thus to figure before the world; so we are fain to restrict our collection, and keep only the comparatively silent members of the Parrot race.

Dr. Russ calls this bird the largest of them all (*der grösste von allen*), but it is much of a size with the Red and Yellow and the Yellow and Blue Macaws, although larger than the Military Macaw, and quite three times the size of the bird to which the name of Illiger was given by Burmeister.

The beak of this species is truly, as Dr. Russ terms it, colossal, jet black, and appears calculated to give a formidable bite, but the creature that owns it, is, at least all the specimens of the species that we have known, extremely gentle, and may be freely handled, even by strangers, which is more than most of the Parrots will permit, speaking much for its intelligence and docility; it is a pity it is so seldom imported, but even in its own country it does not appear to be very common; another incentive to attempt breeding it in captivity.

The Hon. and Rev. F. G. Dutton's account of the Hyacinthine Macaw (Ara hyacinthina).

THE Hyacinthine Macaw deserves to head the list, not only of Macaws but of Parrots, for it is probably the biggest Parrot out. Its colour is a deep, puce blue, not so grey in tinge as the Glaucous Macaw, which is otherwise very like it in size and colour. It has only a very small yellow core at the side of the beak, instead of the bare cheeks of the Red and Blue, and Blue and Yellow Macaws.

The Hyacinthine and Glaucous Macaws differ in a marked manner from the other Macaws, not only in the size of their beak and the portentious strength of their jaws (my Hyacinthine easily bent the wires of one of Groom's indestructible Macaw cages), but also in their disposition. Not that I have anything to say against the disposition of the other Macaws, as will be seen later on, but every Hyacinthine and Glaucous Macaw I have seen has been gentle, and ready to allow any one to handle them. I approach strange Macaws of the other kinds with caution, by no means sure that their tempers may not have been spoilt, or that they may not reserve their affection exclusively for their owners, but I have no fear of the sort with these two Macaws, although I have seen some half dozen, they were all equally good-tempered. They are much less noisy too than the other Macaws, though when they do scream the noise is in proportion to their size. I do not however think that they have the same intelligence as the others, and I am afraid their amiability has something to do with stupidity, as I never came across one that talked. Mine imitated the cackling of a hen to perfection, but it was so occupied with repeating that performance that it appeared to have no time for acquiring any other. The other Macaws did not appear to recognise it as a congener, for they were as afraid of it as if it had been a hawk without the slightest reason, for it was nearly as afraid of them. I did not keep it long, for I like to turn my Macaws out loose, and I was afraid of the mischief this bird's beak might do amongst the garden trees, and as I was offered a good price for it, I let it go. The Hyacinthine Macaw is sufficiently scarce, though not so rare as the Glaucous Macaw: about £10 is the price for either of them.

Before leaving the all blue Macaws, I may mention one other, a *very* rare one, Spix's. This is much smaller than the other two; it is the bantam of the all blue Macaws. It has no naked space at all round the cheeks, the beak and legs are black, and the plumage is a very grey blue. I have only known of one specimen in captivity, that now in the Regent's Park collection. This bird has all the appearance of having been captured as an adult, as its wing appears to have been broken by a shot. Its unsociability therefore says nothing as to what sort of pet a Spix would make. I should think a nice one, if one could get one that had been taken from the nest.

Military Macaw.

MILITARY MACAW

Psittacus militaris, LINN., RUSS.
SYNONYMS: *Ara militaris*, GR.; *Arara militaris*, VLL.;
Psittacus militaris et ambiguus, KL.; *Macrocercus militaris*, JARD., BP.
Sittace militaris, WGL.
GERMAN: *Der Soldaten-Arara.* FRENCH: *L' ara militaire*, VLL.

MORE frequently imported than the preceding species, the Military Macaw is nevertheless not a common bird, and we are at a loss to understand the reason that induced so many writers to bestow upon it a soldierly designation which, in our opinion, should have been reserved for his relative the Red and Blue Macaw; it seemed, however, befitting in the eyes of the great Linnæus, and subsequent authors have tacitly accepted the master's dictum without question.

This bird is an inhabitant of the northern parts of South America, and extends into Central America; it is rather inferior in size to the Hyacinthine Macaw, but is equally noisy and objectionable in a house: it should be treated in the same manner, and is quite as robust. Jardine states that "it is now ascertained to be a native of *Mexico* and Peru, inhabiting the warmer districts of the Andean chain, and attaining to an elevation of about three thousand feet", which is surely a colossal stature for even a Macaw.

As the same author states in another place that the Carolina Conure is the only Parrot that is found in North America, we have no alternative, especially as he couples the Military Macaw with Mexico, than to conclude that he believed the latter country to form a portion of the southern continent of America, which, however, in view of the date at which he wrote, is quite a pardonable error.

This Macaw has a crimson forehead, and a reddish brown chin; the rest of the head, the neck, lesser wing coverts, the back, and all the under parts of the body are green; the rump and upper tail coverts are blue; the tail is scarlet above, with blue tips, and orange yellow beneath, which is also the colour of the under wing surface. The orbits and cheeks are naked, and of a pinky flesh colour, with four narrow stripes, or bands, of a brownish purple colour upon the latter; the irides are composed of a double circle, the outer of which is bright yellow, and the inner greyish green.

It is rather smaller than most of the Macaws, measuring about twenty-nine inches from beak to tail. Wagler asserts that it differs from most of its congeners in many of its habits; in, for instance, that it frequents cultivated fields, where it does much harm to the growing crops, and where, Cockatoo fashion, it places a guard upon the summits of the surrounding trees to give timely warning of approaching danger; which guard is subsequently fed from the crops of some of the party, who disgorge a portion of the spoils they have carried away, for the benefit of their vigilant sentries.

Wagler also states that these birds are in the habit of feeding upon the blossoms of the *Erythinæ*, and *Thibaudiæ;* but whether for the sake of the honey they contain, or for the fleshy substance of the flower itself and the embryo seed-vessel, does not clearly appear from his account.

"It is easily tamed", writes Selby, in his *History of the Psittacidæ*, "and of a docile disposition, but can rarely be taught to articulate more than a few words. It appears to have been a favourite among the ancient Peruvians, as we are told it was frequently presented to the Incas, by their subjects, as an acceptable gift."

Edwards appears to have been the first writer who described this bird, which was figured by him in his *Gleanings of Natural History;* though ignorant when he wrote of its true habitat, he rightly conjectured it to be an American bird.

Writing of this species, Dr. Russ, in his excellent *Handbuch für Vogelliebhaber*, says, "*Heimat Nordwesten Südamerikas und Mittelamerika bis zum Norden Mexicos*", though its occurrence in the latter region seems to us to require confirmation.

Like all the Parrot family, with the exceptions already mentioned, these birds breed in hollow trees, making no nest, properly so called, but laying their eggs, restricted to two throughout this group (*Macrocercinæ*), on the bare wood.

All the Macaws, like the former human inhabitants of their native land, are worshippers of the sun; to judge, that is to say, by the deafening clamour with which they greet the dawn of day. When the great orb of the sun makes its first appearance above the horizon, all of these birds that inhabit the district wake up from their slumbers, and fly, as with one accord, to a common place of rendezvous, generally some decayed patriarch of the forest, whose huge leafless branches seem to furnish them with convenient seats, or rather stands, for the ceremony that is to follow: here, amid the utmost noise and vociferation, they take their stand, facing the rising sun, and display their wings and tails to his genial beams; one might suppose the flock to be engaged in sun-worship; but no, they are merely drying their plumage damped by the heavy dews of night, and their loud conversation is probably nothing more than an expression of their delight at feeling once more dry and warm and comfortable; though, perhaps, they may also be deliberating whither they will proceed to breakfast: the plantation of so and so is very strictly guarded, the crops of some one else have already paid heavy toll, and so on: for when warmed and dried, the Macaws fly off in little parties in search of their favourite food, but meet again in the evening at their accustomed drinking place, and then retire in company to roost for the night.

The hours of feeding, drinking, and reposing are mostly observed with the greatest regularity, so that a person who has watched their habits for a little time, will be able to predict, almost to a minute, their arrival at, or their withdrawal from, a given place which they are known to frequent.

Supposing their toilet to be completed half an hour or so after sunrise, they continue feeding until about 10 a.m., when they fly to the watercourse they frequent to bathe and drink: by the time they have washed and dried themselves again, it is getting on towards noon, and the rays of the sun are descending on the land with almost fiercest power: the Macaws and many other species take shelter from the heat among the densest foliaged trees they can find, and there doze and digest, amid the profoundest stillness until the afternoon, when they pay a second visit to the water and to their feeding-grounds: having satisfied their appetite they retire to the dead tree where they met in the morning, as if to close the day as, apparently, they began it, by some act of homage to the orb of day, and their worship, if such it be, ended, they retire to their usual camping-ground.

During the breeding season, however, the programme of the day is not quite strictly kept; a couple of young Macaws require a good deal of attention, and the parents have to forage far and wide for their support: whether it be the strict habit of this family to have but two young ones at a time, is somewhat doubtful; for it is certain that in captivity, when they do breed, they occasionally have three and four in a brood: we have known as many as seven in one nest, but in that instance they all died, seemingly from inability of the parents to attend to all their wants, for, although they were well supplied with food as a rule, once or twice they did not get exactly what they liked, and when the male, for he is the chief feeder, was relieved of his cares, by the death of the last of the septenniad, he was quite worn by his labours, and scarce looked half his usual size, he was so weak and thin.

"The Macaws", wrote Bechstein, more than a hundred years ago, "are very dear, and are only found in the possession of rich bird-fanciers. Their beautiful plumage forms their principal attraction. In the centre of Germany one costs from fifty to one hundred rix dollars, and in the maritime cities, thirty to forty. They learn to repeat many words, to go and come, and also to obey the least signal from their master: they imitate perfectly the bleating of sheep, the mewing of cats, and the barking of dogs: their custom of drinking only in the evening seems very extraordinary."

Macaws still maintain their price, as we have seen from the sum quoted, as the cost of acquiring one of them, by Dr. Russ: we have seen them drinking in the morning, and at midday as freely as in the evening, so that the habit alluded to by Bechstein must have been peculiar to the individual bird he had under observation when he wrote.

It appears to us that the Military Macaw, if not quite as good a speaker as his Hyacinthine relative, is nevertheless a capable and intelligent bird: and we should like to try and breed him, if only we had a suitable place in which to try the experiment, and we think it would not be difficult with a pair of very tame and healthy birds; but the age of the male would not be of much consequence, as they are of an exceedingly amorous temperament, and retain the fire and passion of youth to quite a considerable age: in the case of the female, however, it is desirable to secure a youthful specimen, as the drain on the system, consequent upon the elaboration of her eggs, would be probably too much for an aged female, who would be almost certain to become egg-bound, or perish in convulsions, as we have unfortunately witnessed more than once.

Given a young female of this, or any other species of Macaw, and a male of any age, provide them with a suitable habitation and appropriate nesting furniture, and we have no doubt they would breed as freely as a pair of Budgerigars or Cockatiels.

We once saw a Red and Yellow male Macaw pair with a female of the Yellow and Blue variety, and we have no doubt that the progeny, had any resulted, would have been capable of reproduction, but the poor birds were, each, chained to a stand, and we do not remember that any eggs were laid; or, if so, they certainly were not incubated.

RED AND BLUE MACAW.

Psittacus macao, LIN., RUSS.

SYNONYMS: *Ara macao*, L.Z.S.; *Psittacus erythrocyaneus*, GSSN.; *Psittacus aracanga*, GML.; *Macrocercus aracanga et macao*, VLL.; *Sittace aracanga*, WGL.; *Sittace macao*, FNSCH.

GERMAN: *Der hellrothe Arara.* FRENCH: *L'ara rouge*, BUFFON.

THIS grand bird is, without a doubt, king of all the Macaws: his gigantic size, forty inches in length, twenty-four of which belong to the tail, his immense beak, and formidable claws, not to say talons, his deafening outcries, his gorgeous plumage, and voracious appetite, are qualifications in right of which he is raised to the *Macrocercian*, if not exactly to the *Psittacidean*, throne, to which not one of his congeners, we imagine, will venture to dispute his title.

He is docile, gentle, even, and teachable to a very great extent. "They learn to repeat many words", says Bechstein, "to go and come, and also to obey the least signal from their master."

But that great master of bird-lore was not fond of the bird nevertheless, for he continues: "I confess, however, that their awkward walk, their heavy movements, and their constant inclination to help themselves along with their beak, added to their great uncleanliness, does not appear very agreeable."

The latter objection might be readily met by feeding the Red and Blue Macaw on a more natural diet than that usually assigned to it, to wit, "bread and milk sop", which was the recognised food of these birds even in Bechstein's time, for, writing on this subject, he says: "In its native country the fruit of the palm tree is its principal food; our fruit it also likes, but white bread soaked in milk agrees with it better; biscuit does not hurt it; but meat, sweetmeats, and other niceties are very injurious: and though at first it does not appear to be injured, it becomes unhealthy, its feathers stand up separate, it pecks and tears them, above all those on the first joint of the pinion, and it even makes holes in different parts of its body."

Poor dyspeptic Macaw, if the picture of your sufferings just quoted had been drawn to-day, instead of a hundred years ago, it could not have been more accurate. What then is the proper food for this species? "Many bird-fanciers", continues our author, "say that the best food for Parrots is simply the crumbs of white bread, well baked, without salt, soaked in water, and then slightly squeezed in the hand. But though this appears to agree with them pretty well, it is however certain that now and then something else ought to be added."

So, indeed, we should say! and Bechstein was of the same opinion, for he goes on to remark: "I have observed that Parrots which are thus fed are very thin, have hardly strength to bear moulting, and sometimes even do not moult at all: in that case they become asthmatic, and die of consumption."

Away then with "sop" and its inevitable concomitants of dirty sour tins and diarrhœa: a Macaw fed on oats, canary seed, maize, and hemp seed, with the addition, now and then, of a lunch biscuit, ripe fruit of all sorts, a slice of carrot, turnip, potato, or even mangel wurzel, will be healthy and his cage clean: on the ordinary diet usually given in this country he is generally unhealthy, and always dirty, which is not by any means the fault of the poor bird, but that of its master, who has it in his power, by treating his Macaw more in conformity with its natural habits, to correct the evils of which he complains in connection with keeping it in captivity.

The appearance of this Macaw, even when caged, or chained to a perch, is such as to at once arrest the attention of the most indifferent beholder; but when seen at liberty, in this country even, actively climbing among the branches of a large tree, or wheeling round and round in the sunshine, like an enormously exaggerated butterfly, he ought to be "a joy for ever", for that he is "a thing of beauty", few, we imagine that have ever seen a perfect specimen of his race, will be ready to deny.

The head, neck, breast, belly, thighs, top of the back, and the upper wing coverts, are brilliant vermillion. The lower part of the back and the rump are light blue. The scapulars and large wing coverts are a mixture of blue, yellow, and green. The naked cheeks are covered with wrinkled whitish skin. The under surface of all the tail feathers is red.

The female bears a general resemblance to her mate, but the colours of her plumage are of a duller shade than his.

This fine bird is a native of South America, abounding in the forests of Brazil and Guiana; Dr. Russ adds "Mexico and Peru", but we doubt its occurrence in the former country; its head-quarters appear to be the Amazon region, which it shares with numerous species of the *Psittticidæ*.

Waterton, writing of the bird under consideration, observes, "Superior in size and beauty to any Parrot of South America, the *Ara* will force you to take your eyes from the rest of animated nature and gaze at him: his commanding strength, the flaming scarlet of his body, the lovely variety of red, yellow, blue, and green in his wings, the extraordinary length of his scarlet and blue tail, seem all to join and demand for him the title of Emperor of all the Parrots", which is a still higher title that we had ventured to confer upon him in the opening sentences of this chapter.

"He is scarce in Demerara", continues our author, "until you reach the confines of the Macoushi country; there he is in vast abundance: he mostly feeds on trees of the palm species. When the concourite trees have ripe fruit upon them, they are covered with this magnificent Parrot. He is not shy or wary; you may take your blowpipe and a quiver of poisoned arrows and kill more than you are able to carry back to your hut."

Let us hope the American naturalist was never guilty of such a piece of wanton cruelty, not to say sacrilege as that he hints at here: he continues, "They are very vociferous, and like the common Parrots, rise up in bodies towards sunset, and fly two and two to their places of rest. It is a grand sight in ornithology to see thousands of *Aras* flying over your head, low enough to let you have a full view of their flaming mantle. The Indians find the flesh very good, and the feathers serve for ornaments in their head-dresses."

These birds make their nests in the holes of trees, which they enlarge and fashion to their liking: there are usually two broods in the season, of two young each time; male and female sit alternately upon the eggs, which are generally two, very rarely three, in number. It is not unusual for the females of this species to lay in captivity, and even to evince a strong desire to incubate, but young ones are rare, and we know of no well authenticated instance in which they have reared their offspring in captivity: seeing, however, that the Red and Blue Macaws that are brought into this country, were taken when quite young from their nest, reared by hand, and are consequently quite tame, there seems no reason why they should not freely breed here, as well as many of their congeners, but in all probability this is because no serious attempt has ever been made to induce them to nest in our aviaries.

Mr. Wiener's account of his experience with this bird and its fellow, which forms the subject of our next chapter, is brief and to the point: "The Macaws I may dismiss with a very few words. I tried a Red and Yellow Macaw, and a Blue and Yellow Macaw. A couple of expensive cages were demolished very quickly, and before a pair of stands could be finished by the maker. The destruction of the hard wood perches and mahogany uprights of their new stands afforded about two days' amusement to the birds, who next peeled off the wall-paper within reach, and gnawed the corner of a billiard-table. This mischief was accompanied by such deafening screams, that a couple of weeks' possession had quite settled my determination to get rid of the magnificent Macaws on any terms, and never to buy one again at any price. Their huge size, brilliant feathers, and loud screams are a very good advertisement for a travelling menagerie, to whom amateurs had better abandon these birds, unless some one would care to construct a wrought-iron in-door aviary (I doubt whether bricks and mortar would be proof against their beaks), to make an attempt at breeding. A pair exhibited some years since at the Crystal Palace were said to have laid eggs in confinement: and as Macaws always arrive in Europe quite tame, it ought to be possible to breed them."

From the foregoing account it would appear that Mr. Wiener was unfortunate in his experience, and that instead of a couple of tame birds, he was imposed upon, and induced to buy two, caught when adult, which, as Bechstein well observes, "are savage and untractable, and would only stun one with their unbearable cries, the faithful interpreters of their different passions."

"Yes", writes Mr. Gedney, who, by the bye, calls this bird "the Military Macaw", "an old trapped Macaw affords plenty of 'raw material' upon which the advocates of 'moral suasion', as a means of taming wild creatures, might very well try their hands. I know one bird that defied every effort made to tame him, and he killed a bull terrier that shared his place in the stables: you could not live in the house with him! Both his wings were broken in this terrific battle, and a pretty spectacle the place presented when the man went as usual to feed him in the morning. There laid poor Tyke dead, with his throat torn open, the bird, covered with blood and almost featherless, stood by, with distended and drooping wings, a perfect scarecrow, shrieking at intervals, either in spite or pain. What was to be done with the creature? Kill him, every one said but the man who looked after the bird; so his belief that the injuries would tame him saved his (the bird's) life: and the cripple was consequently shut up in a pig-stye. His wings got well, the bones growing out of place, but this old savage never abated one atom of his hatred for every one that went near him, and he had ultimately to be poisoned."

We can heartily endorse the following recommendation by the same author:—"Never keep a Macaw in a cage, because, if you do, his gorgeous tail will assuredly be spoiled, and the soiled condition of the cage will inevitably become a nuisance, no matter how great may be the attention bestowed upon its frequent cleansing."

If the owner has not an aviary of sufficient extent and strength to permit of his placing his Macaw in it, and the bird itself is not sufficiently tame to admit of allowing it its freedom, he had better have it fastened by one leg to a stand, by means of a light steel chain: the latter should be attached by means of a ring of sufficient size to admit of its sliding freely up and down to an upright of some strong wood, at the upper extremity of which should be placed a cross bar, the whole taking the shape of the capital letter T; the seed and water tins should be placed at either end of the horizontal bar, and if a well-sanded tray be placed at the bottom of the upright, very little dirt will be made, and the bird be kept in a clean and comfortable condition.

A Macaw thus kept soon becomes very tame, and rarely attempts to bite: as some of these birds, however, are treacherous in their conduct towards children, whom too many have reason to consider their natural enemies, it is as well to caution the young folk against approaching them too nearly.

We have seen tame Macaws as quiet and gentle as any bird can possibly be, and so far from being noisy their voice was very seldom heard, and when utterance was occasionally given to a squeak, rather than a shriek, the note was far from being as shrill and disagreeable as that of the Rosy Cockatoo, for instance, or even the Alexandrine Parrakeet; but no reliable inferences can be drawn from the disposition of individual birds, for they vary in temper, not to say character, as much as men do.

The extreme beauty of a flight of these grand birds is well described in the following extract from Anson's *Voyage*, page 218: writing of a waterfall in the Island of Quibo, he says, "While we were observing it, there came in sight a prodigious flight of Macaws: which hovering over the spot, and often wheeling and playing on the wing about it, afforded a most brilliant appearance, by the glittering of the sun on their variegated plumage: so that some of the spectators cannot refrain from a kind of transport, when they recount the complicated beauties which occurred in this extraordinary waterfall."

It is curious that Latham, when writing of this bird, should fall into the double mistake of saying that "its eggs are spotted as in the Partridge", for the eggs of the latter bird, as is very well known, are of a uniform olive colour, without a spot of any kind, and those of the Macaw are pure white, which we believe, without any exception, is the colour of the eggs of every species of the family, whether found in the old world or the new.

BLUE AND YELLOW MACAW.

BLUE AND YELLOW MACAW.

Psittacus ararauna, RUSS.

SYNONYMS: *Ara ararauna*, GR., BP., *etc.*; *Sittace ararauna*, WGL.; *Macrocercus ararauna*, Auctorum. GERMAN: *Der gelbrüstige blaue Arara*. FRENCH: *Ara bleu et jaune*.

IN the Blue and Yellow Macaw the species last described finds a powerful rival to the throne of Parrotdom, but the bird now under consideration being of less size than the Red and Blue Macaw, the latter must be looked upon as successfully maintaining his claim to superiority, but as there is no very great difference in size between the two species, a couple of inches in length or thereabouts, it would have been a question of great delicacy to decide between the two claimants for imperial honours, had not the magnificent plumage of the Emperor *de facto* rendered it comparatively easy to decide in his favour: bright red and deep blue are more striking colours than dark blue and yellow, so that the possessor of the latter uniform is obliged to fall back, and content himself with second place.

The Blue and Yellow Macaw is of less frequent occurrence in the bird market than its more showily-coated relative, but makes quite as amiable and desirable a pet; of the two, it is, perhaps, the most susceptible of education, speaking with great distinctness, if it does not learn to repeat a very lengthened vocabulary: we say "amiable and desirable pet" advisedly, for we know of few more amusing and interesting birds, providing, that is to say, they are taken in hand from the nest, when they become, as we have said, exceedingly tame.

Their tempers are naturally good, at least all the specimens of the several species of *Macrocercus* with which we have had to do were possessed of a variety of good qualities; they are not, for instance, nearly as excitable as the Cockatoos, or as jealous and spiteful as the Parrots, and they are far more easily tamed than any of the Parrakeets;

but their tempers can be ruined, and too often are; no bird is proof against teasing, which utterly demoralizes them, and soon transforms a naturally amiable and gentle bird into something little better than a fiend.

All the Macaws have large beaks, whence they were not inappropriately named *Macrocerci* by Vieillot; but the bird under notice has, perhaps, the most formidable bill of all its congeners, as it measures no less than three inches and a half from its insertion in the skull to its tip, the under mandible is much shorter, but nevertheless of considerable size, short and massive, and forming a right angle with the upper when shut.

As may be guessed from its name in English this bird is dressed in a mantle of deep blue, the forehead, crown, and rump are of the same colour but with a greenish shade; the tail feathers and primaries are indigo blue, with a violet shade, the cheeks are white, naked, and traversed by four narrow zigzag lines, composed of minute black feathers; the chin is black, but all the rest of the under surface of the body is yellow; the large beak is black, and the feet and legs are dark grey; and the under surface of the tail and wings yellow.

The Blue and Yellow Macaw is a native of Tropical America, but, nevertheless, sufficiently hardy when once fairly acclimatised to resist the cold of our changeable climate, as he has a warm vest of soft down under his robe of blue and yellow feathers. Unlike the preceding species this bird is not found in large flocks, but generally in pairs, which seem to mate for life, and are models of conjugal affection, passing much of their time in caressing and pluming each other: it is a wood-loving bird, too, and seldom approaches the settled districts, so that it has not made itself so many enemies as many members of its race have done.

The eggs, two in number, are placed in a hole in the trunk of a tree, and the young are quite three months before they leave the nest, and nearly three years elapse before they are fully grown; from which circumstance it may be gathered that they are long-lived birds. The sexes share the task of incubation between them, sitting alternately on the eggs and young.

"The dimension and form of their wings", write Selby and Jardine, "and long cuneiform tail, indicate a powerful and vigorous flight, and accordingly we are informed that in this respect they are inferior to none of the tribe, their flight being often at a high elevation, and accompanied with a variety of aerial evolutions, particularly before alighting, which is always upon the summit of the highest trees."

It seems peculiar that almost the first words these birds should learn to speak, and which they invariably pronounce more distinctly than any other, should be the dissyllable "Robert": but so it is, and yet we do not think it forms any part of their natural vocabulary: all the Macaws we have known repeat the word Robert very distinctly, and this seems also to have been the case in the time of the authors just quoted, who observe upon this subject: "A very fine one (Blue and Yellow Macaw) is completely domesticated at Dr. Neill's, Canonmills, near Edinburgh, and allowed the freedom of several apartments; when desirous of being noticed, it calls out "Robert", the name of its earliest master, very distinctly; but it has not acquired more than one other conventional sound."

Several instances are on record in which the Blue and Yellow Macaw has nested and hatched its young in this country and on the continent; and we cannot help thinking that if the birds were more frequently kept in pairs than they are, there would be little or no difficulty in inducing them to breed in an appropriately furnished apartment.

The one great objection to the keeping of all the Macaws is their noise, but this can, by judicious management, be certainly minimised; once a bird has contracted the habit of shrieking, good-bye to peace and quiet, as long as it remains in the house: the obvious remedy being not to permit the bird to acquire the distressing habit, which, once contracted, grows with its growth, and becomes intensified with its increase in years, until at last a parting must take place between bird and owner, and that on the part of the latter in absolute selfdefence.

The Blue and Yellow Macaw is an old inhabitant of the aviary, and instances are on record where it has bred in captivity as far back as the year 1818.

M. Lamouroux, who was the owner of these birds, relates their family achievements in the following terms:—"In four years and a half, from the month of March, 1818, to the end of August, 1822, these birds laid sixty-two eggs, in nineteen broods. Of this number, twenty-five eggs produced young ones, of which ten only died. The others lived, and became perfectly accustomed to the climate. They laid eggs at all seasons; and the broods became more frequent and more productive, in the course of time; and in the end much fewer were lost. The number of eggs in the nest used to vary, six having been together at one time; and these Macaws were seen to bring four young ones at once. These eggs took from twenty to twenty-five days to be hatched, like those of our common hens. Their form was that of a pear, a little flattened, and their length equal to that of a Pigeon's egg. It was only between the fifteenth and five and twentieth day that the young ones became covered with a very thick down; soft, and of a whitish slate-grey. The feathers did not begin to make their appearance until towards the thirtieth day, and took two months to acquire their full growth. It was a dozen or fifteen months before the young arrived to the size of their parents, but their plumage had all its beauty from six months old. At three months old they abandoned the nest, and could eat alone; up to this period they had been fed by their father and mother, which disgorged the food from their bill, in the same manner as Pigeons do."

In all probability this successful rearing was owing to the care which was taken in providing the old birds with a suitable nesting-place, which consisted of "a small barrel, pierced, toward a third of its height, with a hole of about six inches in diameter, and the bottom of which contained a bed of sawdust three inches thick, on which the eggs were laid and hatched."

From the above interesting narrative it will be gathered that no material difficulty is to be apprehended by any aviarist who makes up his mind to breed this species of Macaw in captivity. It is true we are not told where the birds were kept, whether in-doors or in a garden aviary; but as they are quite hardy, when acclimatised, we see no reason to doubt the full success of the attempt if made out of doors in a properly constructed and sheltered building, part of which at least should be open to the air and light.

Nor in the matter of food would there be any difficulty in the way of breeding these birds: for their principal diet is corn of various kinds, part of which should be soaked for them when they have young, and bread should be added as well as fruit, and such vegetables as carrots and potatoes; and as all the Macaws fetch a good price, the experiment, if successful, would be a remunerative one.

In-doors it would even be more likely to succeed than out, for in a well-lighted room, where an equable temperature could be maintained all the year round, there is no doubt these birds would keep on rearing brood after brood from January to December, providing they were supplied with a sufficiency of suitable food.

If the Blue and Yellow Macaw breeds in confinement, we see no reason to apprehend that the other species of the sub-family would not do the same; and the reason that they have not done so is that no pains have been taken with them, for they often pair in captivity, and solitary females not unfrequently lay.

The Hon. and Rev. F. G. Dutton's account of the Blue and Yellow Macaw (Ara ararauna).

WHAT Bechstein could have meant by saying that Blue and Yellow Macaws are not good talkers I do not know; I have had four, two cocks and two hens. The hens did not talk, but the cocks did, and one had a talent for talking, such as I have never met with in any other Parrot. It not only picked up things it heard at once, but always in the tone of the person who said it. It was impossible to doubt whom it was imitating; the only doubt, if it was not mimicking oneself, was, was it the Macaw or the persons themselves? I parted with it, however, first because I could not trust its temper, and secondly because it never would leave a bough it had flown to, if it could help it; what may have startled it in its several flights I know not, but had it been left alone, it would several times have starved to death sooner than take wing again. When therefore it had flown out of the garden, it did not, like the others, return when hungry, but always had to be fetched back, and as this gave considerable trouble when it settled high up in a large tree, I got rid of it. I do not think its temper would have been bad, but I put it in the cage with the Red and Yellow, and it was marital jealousy that made it peck at one. Precisely the same thing happened with the other Blue and Yellow cock bird. He had the best of tempers, any one might do anything they liked with him; but after he was put in the cage with the Red and Yellow, he made efforts to drive people away. Curiously enough neither of the cock birds was a particularly good flyer, but the two hens were as strong on the wing as Hawks, a hurricane would have been nothing to them, and it was magnificent to see the dashes and turns they would make on the wing. They all four had very different characters: "Frank", the good talker, had evidently left his heart for mankind in Brazil. He made distinctions, but he *loved* no one. "Bob", the other cock, was, I think, a little "wanting", anyone might do what they pleased with him, and he would come to a coal-scuttle as soon as to his master. "Harry", as one hen was called, was timid and would only come to myself. I parted with her because she would always settle on just the very leaders of my firs, etc. But I always have regretted having done so. Her wing was cut when she made this her practice, and no doubt had I waited till it was quite grown, she would have returned home from her excursions in a more "convenable" manner. As for "Jenny", the other hen, she was a splendid specimen, as a *bird*, but she was the incarnation of greediness and selfishness; she truly "departed" when I got rid of her, "without being desired."

It only remains to say that the Blue and Yellow Macaw is blue on the back, wings, and tail, golden yellow underneath, green on the forehead, and black under the lower mandible; its eyes, and the eyes of the Red and Yellow, are pearls, and it has the habit of dilating and contracting the pupils like a Paroquet.

They are worth £5 each.

ILLIGER'S MACAW

Psittacus maracana, Russ.
SYNONYMS: *Macrocercus maracana*, VLL.; *Macrocercus Illigeri*, BRMST.; *Ara maracana*, GR.; *Sittace maracana*, FNSCH., etc.
GERMAN: *Der rothrückige Arara*.

THIS rare bird is a native of the southern parts of Brazil, and, although classed with the Macaws by authors, is a much smaller bird than those giants of the Parrot race which we have just described; it is about the same size as Pennant's Parrakeet, though its much longer tail causes it to appear larger.

A dark red forehead, rump, and belly form a marked contrast to the garb of dusky green that clothes the back of this small Macaw, whose wings are blue, tipped with a narrow border of a deeper shade of the same colour; the tail is dark reddish brown, with bluish green tips.

The female resembles her mate.

Two fine specimens are now to be seen in the Parrot House in Regent's Park, but are very seldom to be met with in the dealers' hands, probably because these Macaws have no very marked peculiarities to recommend them to the notice of amateurs.

At the "Zoo" the cage is placed on the upper shelf of the stand that occupies the centre of the Parrot House, where the birds are badly seen, and being out of reach of the majority of the visitors, have not become as familiar with them, as those of their congeners who fill a less exalted position.

It is much to be desired that a new Parrot House were built, on the plan of the Reptile House, where each species of Parrot could have allotted to its use a cage of sufficient dimensions to enable it to use its wings, and to contain a hollow log or tree, in which many of the species would undoubtedly breed; many disputed points in the history of these interesting birds, the Parrots, might then be readily solved, and the poor creatures themselves would enjoy better health, and be seen to better advantage by the public, than at present is the case, confined as they are in miserable cages in which many of them can barely turn themselves round.

Only for its relatively large beak, and the extended patch of naked skin that surrounds the eye, this sombre-looking bird would appear to be more appropriately classed with the *Conures* than the *Macrocerci*, with which, however, the above-mentioned distinctions inseparably connect it; its colours are not striking, nor is the shape of the bird elegant; it appears to be dull and spiritless too, and to lack altogether the vivacity and restlessness that are such conspicuous features of the large Macaw's character. We have seen one of these birds sit motionless for hours, with its beak resting on one of the side bars of its cage—the reader must not, however, too hastily conclude that we remained as motionless as the bird to observe it, for we simply could not do it; what we mean is, that having seen the bird assume the position we have indicated, and looked at it for a few minutes, on our return two hours afterwards, we found it in exactly the same spot from which it had, apparently, never moved all that time: at liberty in an aviary or large room, or even chained to a perch, as most of these birds, the Macaws, are in captivity, it is very likely it would betray quite a different disposition, and be as lively and active as any of the members of its race.

Illiger's Macaw is not a rare bird in its own country, the southern parts of Brazil, but is very seldom imported into this, probably in consequence of its sad-coloured coat, in which all the tints look as if they had been subjected to a process of "washing out", and the colours had either not been "fast", or had been imperfectly restored.

As this is one of the birds we have *not* kept, we cannot say much as to its capacity for acquiring knowledge, including the use, or rather imitation, of articulate speech, and other accomplishments for which so many of its congeners have acquired a world-wide celebrity. Under happier conditions than we have ever known it placed in, Illiger's Macaw might turn out a very different bird, and quite overset the opinion we have, so far, formed of its merits.

The Hon. and Rev. F. G. Dutton's account of Illiger's Macaw (Ara maracana).

I MAY close the account of the Macaws with whose habits I am acquainted with a description of Illiger's Macaw.

This is a miniature Macaw, not as large as an Alexandrine Parrakeet, though having all the characteristics of a Macaw. The prevailing colour is dark green; the cheeks are bare and yellowish, the beak black. The specimens with which I was acquainted were tame, but too noisy for a room. I should think it would make a good talker, though neither of those I saw talked. For any one who wanted to try the experiment of turning out Macaws, but was afraid of the damage that might be done by the larger Macaws, Illiger's Macaw would be an excellent bird to try. Though very good-tempered, I observed in a pair caged together in a dealer's shop, the same jealousy that I have alluded to in my account of the other Macaws: no doubt the jealous one was the cock. They are worth about £2 or £3 the pair.

For food I find Macaws require only bread and milk and hemp seed. I have tried them with Indian corn, but they do not seem very fond of it. They do not eat very much bread and milk after they are grown up, though as nestlings they live entirely on it; the principal food is hemp seed. They are small eaters considering their size.

As to diseases, "Bob" occasionally had a cold in his head, and then he was not let out for a day or two, but with the amount of liberty my Macaws enjoy they naturally do not get ill.

Carolina Parrot, or Conure.

Psittacus Carolinensis, Russ.
Synonyms: *Conurus Carolinensis*, Gr., Lss., etc.;
Psittacus luteocapillus, Vll.; *Psittacus ludovicianus*, Vg.; *Aratinga ludoviciana*, Stph.; *Sittace ludoviciana*, Wgl.;
Centurus Carolinensis, Add.; *Arara Carolinensis*, Sld. et Jrd.
German: *Der Karolina-Sittich.* French: *Perruche à tête aurore*, Buffon.

NO Parrot inhabits so high a northern latitude as the subject of the present notice, whose place in the *Psittacidæan* family has been a matter of much contention with authors, some of whom rank it with the Macaws, and others with the Conures; and probably no other member of the family, with the exception perhaps of the Grey Parrot and the Budgerigar, has occupied so much of the attention of writers, and can boast of so considerable a literature devoted exclusively to itself.

Audubon and Wilson, among American ornithologists, have filled many pages of their works with descriptions of this well-known and, on the whole, popular bird; while Prince Ch. Buonaparte, Wagler and Sir William Jardine have by no means failed in paying it attention. Bechstein and Buffon, amongst many others, have given long accounts of this bird in their writings, not to forget the great Linnæus, who calls it *Psittacus ludovicianus*.

Jardine says: "In length it averages about fourteen inches; in extent of wings twenty-two inches; while the Rev. J. Wood alleges that "the total length of this species is twenty-one inches"—a very considerable difference; the truth lying as nearly as possible midway between the two extremes.

The appearance of the Carolina Conure is exceedingly pleasing, the rich emerald green of the upper plumage is relieved by the vividly orange red of the forehead and cheeks, while the rest of the head and neck are gamboge, and on the shoulder spots of orange red are inter-

mixed with patches of golden yellow: the under surface of the body is yellowish green, and the outer webs of the primaries are bluish green, passing into bright yellow at the base. The inner webs are brown with green tips, the tail feathers are green with the inner webs tinged brownish red. The legs and feet are flesh colour, and the eyes light brown.

As authors disagree on the question of classification, so they are not in accord as to the merits, or demerits, of the species under consideration. Audubon observes: "the woods are best fitted for them, and there the richness of their plumage, their beautiful mode of flight, and even their screams, afford welcome intimation that our darkest forests and most sequestered swamps are not destitute of charms."

"On account of its inability to articulate, and its loud disagreeable screams, it is seldom kept caged", writes Selby in Jardine's *Naturalist's Library;* while Wilson delivers his verdict in favour of the bird, and, as the result of actual experiment, pronounces it to be "docile and sociable, soon becoming perfectly familiar, and capable of imitating the accents of man."

Bechstein remarks that, "its cry is frequent, it is rather wicked, and does not speak; but it well makes up for this by its beauty, the elegance of its form, its graceful movements, and its strong and almost exclusive attachment to its mistress."

"Towards its own kind", says Wilson, "it displays the strongest affection, and if its companions be in danger, it hovers around the spot in loving sympathy."

"When engaged in feeding," continues the same author, "they are easily approached, and numbers killed by one discharge; the work of destruction, however, is not confined to a single shot, for the survivors rise, shriek, fly round for a few minutes, and again alight on the very place of the most imminent danger. The gun is kept at work; eight, ten, or even twenty are killed at every discharge; the living birds, as if conscious of the death of their companions, sweep over their bodies, screaming as loud as ever, but still return to the stack to be shot at, until so few remain alive, that the farmer does not consider it worth his while to spend more of his ammunition."

Writing nearly half a century ago, Audubon observes, "They could be obtained as far up the tributary waters of the Ohio as the great Kesshawa, the Sioto, the heads of the Miami, the mouth of the Maumeee at its junction with lake Erie, on the Illinois river, and sometimes as far north-east as lake Ontario, and along the eastern districts as far as the boundary line between Virginia and Maryland. At the present day (about twenty-five years later) few are to be found higher than Cincinnati, nor is it until you reach the mouth of the Ohio that Parrakeets are met with in considerable numbers. I should think that along the Mississippi there is not now half the number that existed fifteen years ago."

There is no doubt that at the present day (1883-4) their flocks are still further reduced, and that one must travel much further south to find these beautiful but, to the farmer, destructive birds. Their strong attachment to their companions leads to their destruction too, as we gather from Wilson's experience. "Having shot down a number", he says, "some of which were only wounded, the whole flock swept repeatedly round their prostrate companions, and again settled on a low tree within twenty yards of the spot where I stood. At each successive discharge, though showers of them fell, yet the affection of the survivors seemed rather to increase, for after a few circuits round the place they again alighted near me, looking down on their slaughtered companions with such manifest symptoms of sympathy and concern as entirely disarmed me."

"I could not but take notice", continues the same author, "of the remarkable contrast between their elegant manner of flight, and their lame, crawling gait, among the branches. They fly very much like the Wild Pigeon, in close, compact bodies, and with great rapidity, making a loud and outrageous screaming, not unlike that of the Red-headed Woodpecker. Their flight is sometimes in a straight line, but most usually circuitous, making a great variety of elegant and easy serpentine meanders as if for pleasure.

"They are particularly attached to the large sycamores, in the hollows of the trunks and branches of which they generally roost; thirty or forty, and sometimes more, entering at the same hole. Here they cling close to the sides of the tree, holding fast by the claws, and also by the bill. They appear to be fond of sleep, and often retire to their holes during the day, probably to take a regular siesta. They are extremely sociable with and fond of each other, often scratching each other's heads and necks, and always at night nestling as close as possible to each other, preferring at that time a perpendicular position, supported by their beak and claws."

There are, perhaps, few members of the family more susceptible of domestication than the Carolina Parrakeets, providing, that is to say, they have been reared from the nest, or, at least, captured when quite young: adult specimens, however, will breed freely in a large aviary, or bird-room, if provided with suitable nesting accommodation. "A pair", says Dr. Russ, "bred in a small cage in my bird-room, and brought up three, and then five, young ones."

It is better, seeing they are such sociable birds, to keep several pairs together; the greatest difficulty being to distinguish the sexes; the female, however, has the inner webs of the first flight feathers black, and she has rather less of the orange-red markings of the head and face, that are so conspicuous a feature in the male.

The young are entirely green until after the first moult, when the head and face become yellow, and when in this immature state were supposed to belong to a different species, and are described as such by several writers. Latham supposed it to be identical with the Illinois Parrot (*Psittacus pertinax*, Auctorum); but this is a much smaller species found in South and Central America.

When first imported the Carolinas are generally very wild, but a little judicious handling will soon tame them, when their shrill screams will be much less frequently heard; for, like all the *Psittacidæ*, they give free vent to their feelings when alarmed; and, as they are naturally timid, the unknown excites their apprehensions, and their outcries are a natural sequence of their alarm.

These birds are excellent parents, as might indeed be gathered from the intense affection they display for each other, and brood and feed their young with the utmost care and attention.

They are to be fed on canary seed, millet, oats, maize, and breadcrumbs; and, if they are nesting, a portion of the seed should be prepared for them by soaking in cold water for a few hours.

Golden-Crowned Conure, or Half-Moon Parrakeet.

Psittacus aureus, AUCTORUM.

SYNONYMS: *Conurus aureus*, GR., BP., etc.; *Conurus canicularis*, BRMST.; *Conurus brasiliensis*, FNSCH.; *Psittacus regulus*, SHW.; *Psittaca brasiliensis*, BRSS.; *Sittace aurea*, WGL.; *Aratinga aurea*, SPX.; *Aratinga pertinax*, BR., etc., etc. GERMAN: *Der Halbmond-Sittich*.

THIS charming bird is a native of Southern America, and is smaller than the Carolina, which it otherwise resembles, measuring about ten or eleven inches in extreme length.

Dressed in a mantle of vivid green, its forehead is ornamented by a crescentic patch of a deep orange colour, like the crescent on Diana's virgin brow, behind which the feathers assume a greenish blue shade on the top and back of the head, the cheeks and the neck are greenish gold, and the breast and abdomen a dull greenish yellow.

The female can scarcely be distinguished from her mate; but the smaller size and fainter colour of the crescentic or half-moon patch that gives the bird one of its English names, as well as its German designation, is thought to indicate the female; but we must admit that the difference is very slightly marked, and the sexes can only be surely recognised by the actions and demeanour of the birds.

We have no doubt that the Parrakeet described by Bechstein under the name of *Psittacus canicularis* belonged to the species under consideration, although the old German writer called the deep orange yellow of the half-moon markings red, and the greenish blue shade of the top of the head sky-blue, for in other respects the descriptions of Russ and Bechstein agree.

"It is handsome, but does not speak", says the latter author, "and

although a native of South America, is not very delicate or difficult to preserve."

This adaptability of South American species, not only of Parrots, and other birds, but of mammals, and plants and trees, to almost every variety of conditions in which they happen to be placed, is a curious fact to which we have already adverted, though why this should be so is a question not very easily or satisfactorily answered: thus the Passion flower is so far acclimatised that it grows freely out of doors in this country, where it blossoms in profusion, and even matures its golden yellow fruit: and the Green and Red-crested Grey Cardinals brave the cold of our most severe winters with impunity; while many flowers and birds of Southern Europe would perish at the slightest degree of frost, although the mean temperature of their native land is not, by many degrees, as high as that of Brazil and Demerara.

The Half-moon Parrakeets are natives of South America, where they are common and widely diffused: the nest is made in a hollow branch, where the female lays two or three white eggs. In the house they become very familiar, and are very gentle and desirable birds: Dr. Russ relates that a male in his bird-room was so tame that it would fly on to his shoulder, and perch on his finger.

In their native country they do considerable damage to the rice crops, and in captivity are to be fed on canary seed, millet and oats, adding rice in the husk when obtainable: in fact the latter is almost a necessity when the birds first arrive in this country, unless they have been accustomed to our English seeds on their voyage from their native land.

There is no record of their having as yet bred in Europe, at least that we are aware of; but should any of our readers chance either to have successfully bred them, or to know of any one who has done so, we shall take it as a favour if they will kindly communicate the particulars to us through our Publishers, as it adds much to the interest possessed by a species for amateurs if the same has been successfully reproduced in captivity.

The Half-moon, or Golden-crowned Conure is often confounded with the Sun Parrakeet (*Psittacus solstitialis*, Lin.), which is a very different bird, although a native of the same country as the Half-moon: the ground colour of the Sun Parrakeet is bright citron yellow; the face, back, breast, and belly are a yellowish brownish red, the wings green with yellow, black and blue markings, the beak is black, and the eyes reddish yellow, so that the birds can be readily distinguished one from the other: though the latter has occasionally been sold as the female of the former, we do not insinuate with any intention to defraud his customer on the part of the dealer, but simply because the latter know no better.

It is self-evident that dealers and amateurs must look upon birds from a very different point of view: the former regard them as so much merchandise merely, in most cases, and know little or nothing of their habits beyond what is current in the "trade", viewing them from a commercial aspect chiefly, and have little acquaintance with and care less about their habits and requirements than is necessary to maintain them in life, and if possible in health while in their possession—afterwards? *Ma foi, après moi le déluge*, as a Frenchman would say; and we have known of instances in which wrong directions as to treatment were given to customers, so as to ensure the death of the birds, "for the good of trade", within a short time of their purchase by an inexperienced amateur.

However reprehensible such tactics may appear, and undoubtedly are, there is no doubt that they are very frequently had recourse to in the trade; and in fact, in the earlier days of our bird-keeping, we have ourselves been deceived in the manner we have described. In experience lies the safety of the connoisseur against such paltry deception, and he must expect to pay for it, in this connection, as well as in every other. There are respectable and conscientious dealers, it is true, who would scorn to deceive the unwary, and to men of established reputation, with a good character to maintain, we would counsel our readers to repair when about to purchase birds, but there are others who might be much more correctly described as the reverse of honest and fair-dealing.

But even the most upright of dealers is not, necessarily, an authority upon bird-matters, generally the very reverse; he has a certain knowledge of his business, it is true, but then his business is to get rid of his goods as quickly as he can at the least risk of loss to himself; and the highest profit he can command; but of birds we never met with a dealer that had any scientific knowledge whatever: the Latin names he might know by rote in some instances, and repeat glibly enough, more or less correctly; but there his knowledge ended, and when he assumed no more, no harm was done: but when such an individual pretends to know all about everything concerning birds, he can be made the subject of much quiet amusement to the naturalist who interviews him.

We remember once going into the shop of a well-known dealer to inquire for some Saffron Finches, of which we were then in need for the purpose of trying some experiments in the way of mule breeding, and asking him if he had any in stock. Yes, he had a few; what did we want them for? We had heard, we said, that they would breed with Canaries. Certainly, replied the dealer, button-holeing us, after his custom, and looking up confidently into our face, don't you know that is where the Lizards come from? Cinnamons, you mean, we replied, somewhat maliciously we must confess. Of course, replied the dealer, Cinnamons I meant; thus revealing his utter ignorance of the subject: for, of course, if mules were obtainable, which we doubt, between birds of such widely divergent habits as the Saffron Finch and the Canary, although they closely resemble each other in appearance, they would be sterile, as every *hybrid* is: so we had a quiet laugh at our scientific and omniscient friend the dealer, whom we have again and again conducted into similar pitfalls, to his momentary discomfiture; but so overweening is the vanity, and so consummate the self-complacency of the man, that he promptly recovers himself and begins again, as amusingly as ever, to air the knowledge he does not possess, with a pompous assumption of exclusive information that is really "as good as a play."

Well, we have wandered an immense distance from our Half-moons, and must hark back again; observing, in conclusion, that when they are fairly acclimatised they are very hardy and desirable birds, which we can confidently recommend to the notice of amateurs in search of an ornamental and agreeable addition to their collections. They are exceedingly gentle and amiable, and may be caged with the tiniest Astrilds and all the lesser Parrakeets and Love-birds, without fear of danger accruing to the small fry from the really formidable-looking beaks of the Half-moon Parrakeets, which they are very expert in exercising upon anything of a vegetable nature that may chance to come in their way, so that they cannot be kept in any enclosure where it is desired to grow shrubs and trees.

WHITE-EARED CONURE.

Psittacus leucotis, Russ.
SYNONYMS: *Conurus leucotis*, Gr.; *Aratinga ninus*, Spx.;
Sittace leucotis, Wgl.; *Psittacara leucotis*, Vors.; *Micrositlace et Pyrrhura lucotis*, Bp. GERMAN: *Der Weissbäckige Sittich.*

THIS small Conure is probably one of the prettiest members of the sub-family to which it belongs; it is about the size of the Turquoisine, and of equally slim proportions; the greater part of the body is covered with dark green feathers, the face and head are deep brick red, and the cheeks are marked by a white patch; the top of the head is dark brown, and a band of bluish grey encircles the neck, the rump, vent and tail are dark reddish brown.

There is little or no difference between the sexes, and these can only be determined with any degree of certainty by watching a number of the birds together, and securing a pair that seem, by their continual and reciprocal caresses, to have entered into the "holy bonds" of matrimony.

Writing of this bird Mr. Wiener remarks: "This small Conure is only a little larger than the Australian Undulated Parrakeet, and was very rare until a year or two ago, but latterly the birds are frequently offered for sale. I believe no other Conure will afford his owner so much pleasure as this one. A pair I kept for a long time proved very intelligent, lively, and hardy, and were quite free from the destructive mania of other Conures, and never indulged in screaming."

They can scream, however, and that right shrilly, too; but they are not often guilty of such unbecoming conduct; as Mr. Wiener says, they are hardy, witness the length of time several individuals of this species have survived in the Parrot House at the "Zoo."

The native country of this species is Brazil, where, in small flocks of from ten to twenty in number, they make themselves exceedingly objectionable to the cultivators of the soil, by their depredations among

the crops of maize, of which they destroy far more, in apparent wantonness, than they can or do consume. The agriculturists, in revenge for the loss and damage inflicted by them, kill and eat as many of these pretty creatures as they can; and although the old birds are most decidedly tough, the young ones of the year, fattened on the purloined maize, are tender and most excellent *gibier:* it seems a pity, however, to put such charming birds to such a use, for there are plenty of ugly ones to take their place on the Brazilian farmers' tables; and, it seems to us, at the price quoted, at present at all events, £2, and even upwards a piece, it would pay the farmers better to export the White-eared Conures to Europe alive, than to kill and eat them at home; but possibly those excellent individuals are ignorant of the commercial value of their little enemies, or doubtless they would treat them in a different manner, for to eat one of them, at the figure they now command in the bird-market, seems something like eating gold, and the Brazilians, unless vastly changed, have a keen eye to "the main chance", as we remember to have heard from some friends who had had extensive dealings with them.

Mr. Wiener fed his White-eared Conures on "millet, canary, and a little hemp seed, with about a quarter of a sponge-cake daily", which is so excellent and suitable a regimen for them, that we are not surprised to hear him say that upon it his pair "grew daily prettier."

These birds are very pretty and gentle, and soon become very tame; a young male, much petted and attended to, will learn to repeat a few words, and become a delightful companion; they are, however, rather shy with strangers, and should not be unnecessarily alarmed; if they are they will bite, and that sharply, but their doing so is solely the effect of fear.

FESTIVE AMAZON PARROT.

Psittacus festivus, Russ.
SYNONYMS: *Amazona festiva*, SCHLGL.; *Chrysotis festivus*, SWNS., BP., SELB. IN JARD.; *Chrysotis festiva*, GR., FNSCH, etc., etc.
GERMAN: *Die rothrückige Amazone.*

THE Amazons are a numerous family, comprising upwards of twenty species, nearly all of which are inhabitants of the region watered by the great river from which they have derived their common name, and its tributaries, and although natives of the Tropics, or the countries bordering thereon, they are neither delicate, nor difficult to preserve in captivity. Some of them rival the Grey Parrot as linguists and mimics, and all are far hardier than the latter bird.

The species under consideration is not to be confounded with *Psittacus æstivus*, one of its congeners, better known to amateurs, and more frequently imported than itself. All the Amazons are of more or less green colour, and are chiefly distinguished from each other by the markings of the head and face, a few by their size only, and one or two by their altogether different appearance to the rest of their kind.

The subject of this chapter, however, is not of the latter sort, for it is a green bird, darker above, lighter on the lower surface of the body, and has a red frontlet and bridle; above the eyes is a line of blue; the wings are marked with blue and black spots; and the lower part of the back and the rump are red: it is not very frequently imported, nor much of a favourite with amateurs, as it is not a great talker, though susceptible of being rendered very tame.

All the Amazons have short tails, averaging about four inches in length, and the species under consideration has the outer feather on each side margined with blue on its outer aspect, while the remainder are green, marked with red near their bases, excepting the two middle feathers which are wholly green.

Selby declares that "It is docile, and easily tamed, and, being of an imitative disposition, readily learns to pronounce words and sentences with great clearness and precision", which is not at all our experience with the species.

On the other hand, as all these birds differ considerably in capacity and disposition, it is quite likely that an odd specimen, now and again, may be met with that has learned to speak as well and as clearly, as the rest of its compatriots are backward in this respect.

Among the palm groves of its native land, the Amazon feeds luxuriously on fruit, but in captivity is content with a more meagre fare of seed—seeds of various kinds, such as hemp, for which it always shows a predilection, canary seed, maize, and nuts of every description, from the cob-nut of our hedgerows, to the cocoa-nut of its native land; nor does it despise such humble fare as monkey-nuts, and in carrots and beet seems to find a substitute for the oranges and bananas of the tropics.

These birds soon become very tame and domesticated, and if their owner resides in the country, may be permitted to wander at will about the grounds, whence they will return to the house for their food: it is as well, however, not to permit them to ramble far when there is ripe fruit to be picked up in the neighbourhood, as their frugivorous propensities are apt under such circumstances to exert themselves with a degree of intensity that cannot fail to prove injurious to the gardener: at other times they may have full liberty, which we have never known them abuse by straying away altogether from their home.

It is curious that a pair of these birds will sometimes converse with each other in their acquired language, but such is nevertheless the fact. Some years ago a friend of ours had a pair of Amazons, though we cannot now say to what particular species they belonged, that used to talk to each other in Portuguese, which they had no doubt learned before their importation into this country. The effect was decidedly peculiar, sitting one in a pear-tree in the garden, and the other in a clump of hawthorn near the dining-room window, they regularly answered each other, and occasionally sang and laughed aloud, so that they were often taken for human beings by persons who had not seen them, and only heard the sound of their voices in the garden. What they were talking about, we regret we cannot say, as we are not acquainted with the language in which their conversation was carried on.

DUFRESNE'S AMAZON

Psittacus Dufresnei, Russ.
SYNONYM: *Chrysotis Dufresnei,* LVLL.   GERMAN: *Die Granada-Amazone.*

THIS uncommon bird is even less frequently seen in captivity than the species we have just described; there is a fine example in the London Zoological Gardens, but it is very seldom imported by the continental dealers. It is a native of the Brazils, inhabiting pretty nearly the same region as the Festive Parrot, from which it is distinguished by a brownish yellow eye-streak, or bridle, and blue cheeks and chin: the frontlet, as in the latter species, is bright red; the wings and tail green, streaked and tipped with blue, red, and black, and the upper tail coverts a brilliant crimson.

In size this bird somewhat exceeds the common Amazon Parrot, measuring from fifteen to sixteen inches in extreme length, while that of the ordinary species is about twelve inches; the wings are large and well feathered, and the bird capable of strong and prolonged flight.

On the continent it is known as the Granada Amazon, as well as by the name that heads this chapter, which was given it in honour of a French naturalist, by Le Vaillant, his countryman, the well-known traveller.

Notwithstanding the general family likeness to each other borne by all the Amazons, it is probable that they belong to different species, rather than that they are local varieties of one or two; a point, however, which can only be cleared up satisfactorily when amateurs, or Zoological Societies, try the experiment of cross-breeding with some of them, and then ascertaining whether the young so produced are or are not capable of reproducing their kind. We are ourselves inclined to believe that, in some instances at least, this would be really the case, as happens with several kinds of Pheasants, formerly believed to be distinct species, but now ascertained to be merely geographical

DU RESNE'S AMAZON.

variations of the same: and, in point of fact, there is more resemblance between the Amazons than there is between the Golden and the Amherst Pheasants, the cross-bred progeny of which has been satisfactorily proved to be fruitful. Again, some of these Parrots are so nearly alike, that it is difficult to believe they are not varieties of the same species, and certainly at first sight at all events, there appears to be much more dissimilarity between a Cochin China and a Game Fowl, than between the Blue-fronted Amazon and its relation with the Yellow Bridle, yet the offspring of the two fowls mentioned above is undoubtedly fruitful, proving the parents to belong to one and the same species, for hybrids are incapable of generation.

It is true that instances have been recorded, at various times, of mules having given birth to offspring, but there is a strong element of doubt in every such case that has been brought under our notice, so strong indeed, that we are of opinion that no verdict is possible, save the convenient Scotch alternative to guilty, or not guilty?, viz. not proven.

The solution of the problem with regard to the unity of the Amazons, is not very difficult, as solitary females not unfrequently lay eggs in captivity, and would, there is no doubt, under favourable circumstances, pair, and probably bring up their young.

Few private individuals have the accommodation or the leisure, not to mention the means, necessary to institute, and successfully carry out the experiments necessary to a complete solution of this and similar problems; but the Zoological Societies of this country and on the continent are in possession of the requisite conditions, and to them we must look for the elucidation of many an ornithological puzzle; most of them, however, will not move in a new direction without a considerable amount of external pressure, which a concerted movement on the part of amateurs is capable of being brought to bear upon them, and we hope, in the matter of the *Psittacidæ*, will very soon be exerted.

The Hon. and Rev. F. G. Dutton's account of Dufresne's Amazon.

This is a handsome bird. He is about the size of the Common Amazon. The prevailing colour is an even green, but the upper tail coverts are brilliant crimson. He has a brownish line of feathers from eye to eye over the beak, which is dark horn colour.

They are not very attractive cage-birds. I have never seen a specimen that talked. They are rather quiet and dull. Their food is the same as that of the Common Amazon.

BLUE-FRONTED AMAZON.

Psittacus Amazonicus, Russ.
SYNONYMS: *Chrysotis æstiva*, Swns.; *Amazonicus fronte lutei et Psittacus brasiliensis cyanocephalus*, Brss.;
Psittacus æstivus, Gml., Lthm.; *Psittacus aouron*, Shw.; *Psittacus Amazonus et Amazona Amazonica*, Schlgl.; *Psittacus agilis*, Ltd., etc. etc.
GERMAN: *Der Amazona-Papagei*, Russ; *Der gemeine Amazonenpapagei*, Bechst.

THIS many-named bird is one of those most frequently seen in captivity in this country, as well as on the continent of Europe: so well known that Bechstein, writing towards the close of the last century could say of it: "This species is imported in so great numbers that it is found at every bird-seller's, and is one of the cheapest."

"This bird," he continues, "is common in the hottest parts of America, learns to speak, is very docile, sociable, and requires only common attention."

"It is frequently brought to Europe," wrote Selby, "on account of its colloquial powers, and known, like others of its congeners, by the common appellation of Green Parrot."

A specimen recently in the possession of a friend of ours was known to be at least sixty-seven years old, having been in one family for nearly forty years, and to the last retained a remarkable degree of health and vigour: it succumbed to an acute attack of inflammation after a couple of days' illness, induced by exposure for a few minutes to a draught.

This long-lived bird was presumably a female, for it never learned to speak, beyond repeating, in a low whispering tone of voice, a few short words, such as 'Polly', 'Kiss me', and so on. A *post-mortem* examination was not permitted, so the question of the poor thing's

sex was never determined; but, as the males of this species are usually very fluent talkers, the probability is that it was a female.

There are many recorded instances of these birds laying eggs in captivity, but none, with which we are acquainted, of their having produced young.

"Of all known animals, there are none so calculated to attract the attention and admiration of man, as those which appear to approximate to his own nature, and to partake of some of the attributes of humanity. This is the case with the apes among the mammalia, and the Parrots in the class of birds. Both exhibit some of the physical peculiarities of man, and both present a very striking analogy with each other.

"The ape, from his external form, so like the human, his gestures and gait, the rude resemblance of his face to that of man, from the analogous arrangement of all his organs with ours, has been regarded as a species of imperfect and wild man. Had he received the gift of speech, like the Parrot, he would have passed for a genuine man in the eyes of the multitude, who judge always rather from external appearances than calm and reflective examination. The Parrot is in the order of birds what the ape is in that of viviparous quadrupeds. It would appear, on first view, to be still more closely connected with us, than the latter, because the communion of speech is more intimate than that of mere sign and gesture. Besides, speech is the expression of thought, while gesture is nothing but the demonstration of physical wants. The latter is altogether corporeal, the former appertains to the mind.

"We must not, however, consider the articulated voice of the Parrot as a proof of the superiority of his intelligence over that of other animals, or of its analogy with our own. It is certainly true, that the Parrots exhibit the most perfect brain to be found among any of the feathered race. The anterior lobes of its hemispheres are more prolonged than they are in rapacious birds, and the encephalon is wider and more flatted than long; but as to the intelligence of the bird, compared with ours, it can only be considered that there is a point of contact between them, as it were, but no resemblance. The Parrot's imitation seems purely mechanical; it articulates words, indeed, but this cannot be deemed a true language. In the same manner as an air is taught to a Linnet with a bird-organ, so a word is taught to a Parrot, which he repeats without knowing wherefore. He does not comprehend its signification, and though he may repeat it on certain occasions, because he has learned it, he sees no reason for doing so like man. He utters, indifferently, a prayer or an insult, and those involuntary substitutions, which really prove his want of intelligence, pass, with unreflecting persons, for a mark of wit, of irony, or of some other quality of mind of which the animal is utterly destitute and incapable of acquiring.

"There are two kinds of imitations: one which is altogether physical, and dependant on similitude of organization; the other, the fruit of reflection, volition and intelligence; the first is possessed by the ape and the Parrot—the second by man alone; one requires nothing but memory, and an aptitude of organic functions—the other demands a profound study, like that of comedians and tragedians. A more imitation of the exterior, such as a brute can give, is insufficient. The mind and soul must be moulded, as it were, on the model imitated; this requires a certain equiponderance of mental faculties, which cannot exist between man and a brute of any species.

"The imitations of which we have been speaking differ again in an essential point. It is thus: the imitation which the animal can acquire being totally physical, perishes with the individual.

"Many stories have been told, and repeated *usque ad nauseam*, of the marvellous deeds of these birds supposed to be consequent on their mental faculties; indeed, most persons are in possession of anecdotes, more or less wonderful, of particular individuals, which have fallen under their own observation, or that of their friends—anecdotes, which too often increase by repetition, till the true extent and character of the original facts are lost. Parrots will certainly sometimes repeat a word or a sentence, which circumstances may render particularly apt and applicable, as monkeys will sometimes use a gesture or an action strikingly human in its appearance; but a very slight acquaintance with these animals will convince any reasonable person that these imitative or mechanical qualities are not to be attributed to superior reason or sagacity; and, as much has been already said upon the subject, we shall not subjoin any repetition of thrice-told tales, or search for others of a similar character, which, however amusing, may be considered as destitute of instruction, and of equivocal veracity."—CUVIER, *Règne Animal*.

Without endorsing the whole of the remarks just quoted, we may observe that we have possessed Parrots, and known others, that seemed to attach a certain significance to certain sounds: thus, an old Cockatoo of ours never called for "Potato!" except when he saw us sit down to dinner; and never said "Oh you're a beauty!" but to a child; and again when he was angry he would exclaim, "Oh you bad Polly!" or "Oh you rascal!" never once making use of the many endearing expressions he was so lavish of at other times, which would tend to show that he had at least some idea of the use of words, some comprehension of the import of what he was saying.

That a Parrot is possessed of wit, or irony, is incredible, but that he does attach a meaning to certain words is, we think, incontestible. It must be remembered that some of these birds are much more intelligent than others, and in this respect the males appear to excel the females, which are usually incapable of learning much. To this rule there are certainly exceptions, but these are few and far between; a good talking bird may safely be set down as a male, and a quiet, silent, meditative one as a female. Speech, however, being "silvern", and silence "golden", it follows that the lady birds have the advantage over their mates in the matter of true intelligence: *vivent les dames!*

The Hon. and Rev. F. G. Dutton's account of the Blue-fronted Amazon (Chrysotis æstiva).

THIS, next to the Grey Parrot, is about the commonest Parrot kept. It never, however, fetches as low a price. One may buy a Grey Parrot for fifteen shillings, but Amazons are generally about twenty-five shillings. It is one of the best to keep, as it learns quickly to talk, and when it becomes a good talker, gives up screaming. I find them a better-tempered Parrot, as a rule, than the Grey. They are more apt at imitating sounds than the Grey Parrot. The Grey does not pick up laughing, crying, and such like sounds as the Amazon does. The Amazon, too, has a special power of giving the idea of a conversation. You hear no word distinctly, but you would certainly say two people were talking together. An Amazon, too, talks much more freely before strangers than a Grey; and certainly one that really talks well is to be preferred to a Grey for that reason. But though I have had Amazons which ceased to scream when they had learnt to talk, they are not sure to give it up, and if a Grey becomes a talker, it almost always does. In that respect the Grey is the better bird to keep. The Amazon is unlike the Grey in this: when people notice it, it will spread its tail and wings, and contract its pupils, like a Bengal Parrakeet; I have never seen a Grey Parrot do anything of the sort. And, like the Bengal, the Amazon sometimes gives you a nip when in this state of excitement.

They are remarkably hardy birds, and can easily be taught to fly about loose, and find their way home. But I do not let my Amazons out, if they are good talkers. Giving them their liberty makes them forget their talking and return to screaming.

This leads me to say that in any attempt to acclimatise Parrots, such as those attempted by Mr. Buxton at Northrepps and in Surrey, we should ask ourselves what end we propose to gain. There seems to be no reason why some of the species should not really be acclimatised, that is become really wild birds. The Cockatoos, I conceive, might. But would it be desirable that they should? Do we want Cockatoos added to our native birds? I imagine the farmers would find them a great nuisance. Of course, if we only want to adorn our own grounds with various exotic species, we have only to choose the sorts we admire most. But there are one or two species which would be a great addition to our native Fauna, and probably annoy no one. It astonishes me that greater efforts have not been made to acclimatise the Budgerigar. There have been times when this lovely bird has been brought over in such quantities, that it has been sold for two shillings and ninepence a pair. Those people who have aviaries might do something. The difficulty is with the first letting out. A bird let out for the first time has to be followed up. One must not only know where it is, but it must know where its master is. After it has once or twice come back to its cage, the difficulty is over; and this is what makes it so easy to train Cockatoos, Macaws, Amazons, and Grey Parrots to liberty. They are easily seen and heard, and if they have flown too far, are striking to strangers, so that one is quickly put on their track. But Budgerigars are so small, that they are easily lost sight of. Acclimatising them might be done in one of two ways. Either the experiment may be tried with single specimens which are very tame, so tame as to fly on to the hand; or with great numbers which have been accustomed for some time to be fed near the wire of an out-door aviary. One might trust to some of the number coming back to feed, and so bringing the others, till they had learned to find their own food. Of course the experiment will best succeed in a very strictly preserved country, where all the proprietors are friends. The Hawks will be shot, and a word to the gamekeepers will save the Parrakeets from a like fate. Blue Mountain Lories, I think, might also be acclimatised, and with advantage. It would be a great additional beauty to our woods, were so splendid a bird to be seen amongst them. The *Platyceri*, I am afraid, are too shy and timid to be likely birds for successful experiments; but Amazons and Greys are not at all birds to try. Their whole recommendation is for strictly cage birds, and they can always be bought so cheaply, that there is but little advantage to be gained from their acclimatisation. They are not particularly beautiful, and their natural cries are far from being a sound to be added to those of our woods.

The food of the Amazons consists of bread and water, hemp, canary and millet seed, half-boiled Indian corn, nuts and fruit.

As to their diseases, they are sometimes taken with sickness, and apparent indigestion. I have always found the best remedy for this state of things to be a little carbonate of soda added to their water. A few dried red chilies, now and then, are very good for them. They should always have a piece of wood to gnaw—fir or larch is best, if procurable.

Double-Fronted

or Le Vaillant's Amazon.

Psittacus Le Vaillanti, Russ.
Chrysotis Le Vaillanti, Gr.; *Psittacus ochrocephalus*, Lchst.
GERMAN: *Der doppelte Gelbkopf*, Russ.

THIS remarkably fine bird rivals the world-renowned Grey Parrot as a mimic, and volumes might be filled with anecdotes, more or less authentic, of its performances in this respect, but *cui bono?* It is found in the northern portions of South America, including Guiana, Surinam, and Venezuela.

A glance at the illustration will give the reader a more perfect idea of the bird, than ten pages of letter-press could do. It is one of the largest of the *Chrysotis* sub-family of the *Psittacidæ*, considerably exceeding in size the Grey Parrot, and approaching that of the Cockatoo, but its short tail, barely four inches in length, gives it the appearance of being a smaller bird than it really is.

The dense forests of its native land abound in nuts and fruits of many and various kinds upon which the Amazon Parrots subsist for the most part, although not averse to maize, for the sake of which they often make incursions on the cultivated grounds, and pay for their thievish propensities with their liberty, for the cultivators catch them with limed twigs, and forthwith sell them into hopeless slavery. Such Parrots, however, rarely become absolutely tame, and never make good talkers; to educate, thoroughly, one of these creatures it must be brought up from the nest by hand, and by the time it can eat alone it will not only be perfectly familiar with its foster-parent; but have probably learned to repeat some words, if not a sentence or two.

Like most of the productions of Tropical South America, the Double-fronted Amazon is perfectly hardy, and would certainly become accli-

matised in our woods were it not for the unhappy propensity, common, alas! to every class of society in these civilized (?) islands, to shoot and destroy a strange bird the moment it ventures to put in an appearance; so that experiments, of the highest interest to naturalists, have been utterly frustrated, though not undertaken without considerable expense, simply on account of this wanton and barbarous habit of "potting" anything strange and unusual in the shape of a feathered fowl: indeed so strong is this inherited propensity in some people that more than one stranger has been hooted, and even stoned in a remote village, for no other reason in the world than because he was a stranger: and to the same cause is doubtless referrible the irresistible propensity common to the entire fair sex, of picking to pieces, metaphorically, a sister whom they chance to see for the first time: but after all, our boasted nineteenth century civilization and refinement is a very thin veneer, strain it but a little, and it forthwith cracks, and shows, unpleasantly enough, the disagreeable savagery that lies hidden close beneath. Well, probably our coat of paint, or gilding, or whatever we like to call it, will grow thicker in due time, and become a real thing, and then we shall cease to stone and stare at a stranger, whether male or female, and to kill strange birds.

To return to our Double-fronted friend, of which a very good example, from which our illustration is taken, exists in the Zoological Gardens in the Regent's Park, where it has lived, without water, for several years. Now, although we are perfectly well aware, as a writer in that grandmotherly Review, *The Saturday*, recently pointed out, that Parrots can exist without drinking, we maintain that it is unnatural for them to do so; and granted that in their wild state some of them, *Psittacus erithacus* for instance, seldom resort to the water-courses, it should be remembered that in the countries where these birds are found the dew falls very heavily, and the leaves during the night are saturated with moisture, which, on more than one occasion, we have seen birds, Parrots included, eagerly sucking before they left their roosting-places to seek their food in their accustomed haunts: but in captivity, where, as often as not, their food consists of dry seed, they have no opportunity of drinking dew, and require to be supplied with water, if they are to be kept in health. It is no answer to say that they can live without drinking; the question is, does it make the poor things suffer? and there can, we think, be no doubt that it does. In no other part of the world with which we are acquainted does the absurd custom prevail, and when we have mentioned it to foreigners, our statement has been received with an astonishment bordering on incredulity.

On the authority of an observer (Beobachter) in Venezuela, Dr. Russ, in his excellent *Handbook*, remarks: "Of all Parrots found in this region the one under consideration learns to speak the most readily and distinctly"; and this appears to be the general verdict with Le Vaillant's Amazon, which English dealers commonly speak of as the Double-fronted; why, it would be rather difficult to say, seeing that the face of the bird is all of one colour, namely, a pale yellow; the shoulders are red, and the rest of the body green, darker above, and of a lighter shade on the under surface.

The Hon. and Rev. F. G. Dutton's account of the Le Vaillant's Amazon (Chrysotis Le Vaillanti).

ACCORDING to my experience, the cleverest and most accomplished birds are found in this family, though their powers of talking vary a good deal with individuals. I have had three, varying from one, quite the cleverest and most charming Parrot I ever had, to another which was quite as distinguished for its want of cleverness and amiability.

My clever one I bought in Brest, from an old couple. I suppose the man had been a sailor: the home appears to have been a place where strength rather than choiceness of language was the rule. It would have been impossible for me to have kept the bird, had it not been French. Its language was enough to make one's hair stand on end. But it sang several songs, did the soldiers' exercises, and had many other phrases, all of which it repeated whenever I wished it. I imagine it would always have done so for the person who fed it; otherwise it would not talk for all the world. It would always talk for any labourer or any man with a gruff voice, be he French or English, and swear too. It laughed with as vulgar a laugh as one can well conceive. I gave one hundred francs for it, but it was well worth three hundred. I suppose its talent was too much for it, for it died about two years or so after I bought it, of cancer on the brain.

Before this I had been very much taken with one I had seen in some lodgings in London, and so procured one for myself from Liverpool. I bought it quite young, and it only proved a moderate talker while I had it. I ought to have said that my French one was absolutely good-tempered: I could do anything with it. I had to hold it while it underwent a cruel operation in the hope of cure. Of course during the operation I held it in a towel. But the moment it was over, it had not the least resentment.

The Liverpool one was not so good-tempered, but I do not know that I kept it long enough to try. I doubt if I was patient enough:

I ought to have kept it longer, for I rather think these birds take several years to get to their full powers.

The last one I had I bought partly from pity, and partly in memory of "Cocot", the French one. It had lost an eye, but I was assured it had been a fine talker. However it never uttered a word with me, and was incurably morose. I made it sit on my finger, and taught it not to bite me, but it would dash at anybody who went near the cage. I eventually lost it by turning it out. It never would eat anything but hemp seed, and this of course stopped my having any chance of taming it by giving it luxuries. Few Parrots resist sponge-cake or pea-nuts, but this refused everything but hemp seed. I expect it had been caught old. If I were buying a "Double-fronted" Amazon, for that is the name by which dealers call them, I should try and buy a cock bird; and for that purpose I should choose one with as yellow a head as possible. The amount of yellow may not entirely be a question of sex. I dare say as the birds get older they get a greater amount of yellow on the head, but I always think the male bird has a yellower head than the female, and for talking I would always rather have a male bird, just as I would for singing.

I have come to the conclusion that as a pet for a *cage* bird, for my experience is confined to cages, as I have no aviary, the best Parrot is a Jardine's Parrot (*Pœocephalus Gulielmi*), the next, a Double-fronted Amazon, and the third, a Grey.

RED-VENTED PARROT.

Psittacus menstruus, Russ.

SYNONYMS: *Pionus menstruus*, GR.; *Pionias menstruus*, WGLR., FNSCH., etc., etc. GERMAN: *Der blauköpfige Portoriko-papagei*.

THIS is a very handsome bird, with a very unbecoming name, but naturalists are not over particular when an appellation suits them; the general colour is dark grass green, the upper wing coverts are greenish olive, with a bronzed reflection in certain lights, and the lower green; the head and neck are violet blue, the ears are black, the tail dark green, the lower tail coverts purple, and the vent blood-red.

Male and female are exactly alike in general appearance; it is another Brazilian species, but extends into La Plata.

Wagler separated this bird from the genus *Psittacus*, and formed it, with some other species, into a separate genus which he distinguished by the term *Pionus*, but it seems to us a pity to multiply genera on such slight, and, to our mind, inadequate grounds. The characters of the former genus, *Psittacus*, according to Wagler, are—Bill strong, proportionate, the upper mandible with the culmen slightly narrowed, the tip, with its under surface, rough with elevated ridges, strongly toothed or emarginate, under mandible slightly compressed, with the cutting edges sinuate.

In the latter genus, *Pionus*, the characters given are—The bill large, the culmen biangulate, the tomiæ sinuate, but not distinctly toothed. Differences, surely, scarce sufficient to warrant the creation of two genera, where the general appearance, and, especially the habits, no less than the habitat of both are, in almost every instance, identical, or at best so slightly divergent as to point to generic unity, and concentration being the order of the day in other circles, we have no hesitation in including the subject of the present notice with the rest of the short-tailed Parrots, of which the Amazons and the Grey are the most familiar examples.

The Red-vented Parrot is not very commonly seen in this country, and consequently commands a high price, which is, in our opinion at least, quite out of proportion to its merits, for, although undeniably handsome, and as a rule very tame, it makes, at best, a very poor talker, but different specimens vary a great deal in intelligence and capacity for acquiring human speech.

The Hon. and Rev. F. G. Dutton's account of the Red-vented Parrot (Pionus menstruus).

Why this bird should be called the Red-vented Parrot, zoologists only know! It is true it has a red vent, but that is quite the least noticeable point about it. The distinguishing part of its plumage is its violet head and neck, and it is much more aptly named by the French, *Le Perroquet à camail bleu*.

It is much smaller than the Amazon, but about the same size as the White-fronted Amazon. The beak is horn colour, with a red spot on each side of the upper mandible. The head, neck, and part of the breast are bluish violet, and the feathers on the belly are tipped with blue. The bird is green on the back, but the wings are a yellow green. The vent, as the name implies, is tinged with red. The tail is green, with red at the root of the inner web of the first three feathers.

Bechstein says the bird comes from Guiana, does not talk, and is very tame and gentle. These remarks I can endorse. All the *Pionuses* I have seen have been very tame and gentle. Mine took a strong dislike to one man, but the rest of the world could do what they pleased with it. It was very quiet, never screamed, but never learnt anything. It was a stupid bird. It never made any distinction except in the one case I have mentioned. It had no more affection for the person who fed it than it had for any one else. When it flew, it settled on the ground and remained where it lit. It had no idea of coming home again. It was lost owing to this. It flew away, and we did not see where it lit. We could not find it, and it probably fell a prey to a fox. It never washed. Its food is the same as for an Amazon.

DUSKY OR VIOLET PARROT.

Psittacus violaceus, Russ.
SYNONYMS: *Pionus violaceus*, FNSCH.; *Pionus purpureus*, WGL.;
Pionias violaceus, BP. GERMAN: *Der Veilchenpapagei*.

THIS is a very handsome bird, about the size of a small Grey Parrot (*P. erithacus*): the plumage is blue and violet of different shades, with some brownish markings on the face, and a very narrow eye-streak or bridle of a deep red colour; the primaries are black, bordered with deep blue on the outer edges, the tail is purple, and the under surface of the body has a brownish tinge washed with purple.

It is an active and lively bird, and looks as if it might become a talker, if taken in hand when young: it is a native of Brazil, and, as usually happens with the birds of that country, quite hardy, and easily kept on a diet of seed.

It is very seldom imported, and we do not recollect ever having met with a specimen in the dealers' shops, but the London Zoological Society possess an individual of this species that has been in their possession for some time, a proof, if any be wanting, that it is not delicate, or difficult to keep; for few of the *Psittacidæ* attain to a great age in the "Parrot House," always, of course, excepting the veteran Vasa, who has lived there since 1830! and still appears to be in the enjoyment of a fair amount of health and vigour.

The Violet Parrot, like the Red-vented, is usually placed in Wagler's genus *Pionus*, but is, nevertheless, a thorough *Psittacus* in shape and habit, and is an extremely lively and interesting bird, and were we writing a scientific instead of a familiar history of these birds, we should have no hesitation in restoring it to its place in the latter genus, as instituted by Kuhl, who divided the *Psittacidæ* into five groups or sections, namely, *Arainæ*, the Macaws; *Plyctolophinæ*, the

Cockatoos; *Psittacinæ*, the Parrots proper; *Conurinæ*, the Long-tailed Parrakeets; and *Psittacalinæ*, or Dwarf Parrots—an arrangement which has no doubt its advantages, but which is, nevertheless, somewhat too condensed, especially with regard to the fourth division, which includes in it such widely differing birds as the Ring-necked Bengal Parrakeet, and the Green Leek of the Australian colonists.

The Hon. and Rev. F. G. Dutton's account of the Dusky Pionus (Pionus violaceus).

This bird is the same size as the Red-vented Pionus. Its feathers are dark grey, almost black, tinged with violet. Precisely the same remark applies to it that apply to the former bird. The one I kept was tame and gentle, but showed no disposition to learn. I had it as a nestling, and I am not sure but that it did know me apart from others. At any rate it made an incipient noise when I was in the room. I had to feed it, and I suppose it found it less trouble to be stuffed by me than to feed itself—at any rate its clamour for me to come and feed it was incessant.

I saw the other day a *Pionus senilis*, which is very like the Dusky Pionus, only that it has a white forehead, which could whisper "Pretty Polly" in a very small voice. This is the only case I have known of a Pionus talking. Its price was £2. It was very tame, and when I say that I love a tame bird, *and did not buy it*, I give my opinion of Pionuses more plainly than if I took pages in which to state it. They are really too dull.

INDEX.

	Vol.	Page.
Alexandrine Parrakeet	i	51
Amazon, Blue-Fronted	ii	98
Double-Fronted	ii	104
Dufresne's	ii	96
Le Vaillant's	ii	104
Parrot, Festive	ii	94
Barraband's Parrakeet	i	67
Beautiful Parrakeet	ii	29
Bengal Parrakeet	i	57
Blood-Rumped Parrakeet	ii	23
Blossom-Headed Parrakeet	i	63
Blue and Yellow Macaw	ii	75
Blue Bonnet Parrakeet	ii	21
Blue-Fronted Amazon	ii	98
Blue Mountain Lory	i	39
Blue-Winged Parrakeet	ii	37
Bourke's Parrakeet	i	107
Broadtail, Yellow-Rumped	ii	8
Budgerigar	i	111
Carolina Conure	ii	84
Carolina Parrot	ii	84
Cockatiel	i	33
Cockatoo, Goffin's	i	1
Great **White-Crested**	i	7
Leadbeater's	i	13
Lesser Lemon-Crested	i	17
Rose-Breasted	i	21
Slender-Billed	i	27
Conure, Carolina	ii	84
Golden-Crowned	ii	88
White-Eared	ii	92
Double-Fronted Amazon	ii	104
Dufresne's Amazon	ii	96
Dusky Parrot	ii	110
Elegant Parrakeet	i	83
Festive Amazon **Parrot**	ii	94
Goffin's Cockatoo	i	1
Golden-Crowned Conure	ii	88
Golden-Crowned Parrakeet	ii	16
Great White-Crested Cockatoo	i	7
Green Ground Parrot	i	121
Grey-Headed Love-Bird	i	135
Grey Parrot	ii	41
Half-Moon Parrakeet	ii	88

	Vol.	Page.
Hyacinthine **Macaw**	ii	61
Illiger's Macaw	ii	81
Javan Parrakeet	i	87
King Parrakeet	i	93
King Parrot	i	93
Leadbeater's Cockatoo	i	13
Lesser Lemon-Crested **Cockatoo**	i	17
Le Vaillant's Amazon	ii	104
Lorikeet, Swift	ii	33
Lory, Blue Mountain	i	39
Purple-Capped	i	45
Love-Bird, Grey-Headed	i	135
Madagascar	i	135
Rosy-Faced	i	129
West African	i	125
Macaw, Blue and Yellow	ii	75
Hyacinthine	ii	61
Illiger's	ii	81
Military	ii	65
Red and Blue	ii	69
Madagascar Love-Bird	i	135
Many-Coloured Parrakeet	ii	26
Mealy Rosella	ii	6
Military Macaw	ii	65
New Zealand Parrakeet	ii	13
Orange-Bellied Parrakeet	i	103
Pale-Headed Rosella	ii	6
Paradise Parrakeet	ii	29
Parrakeet, Alexandrine	i	51
Barraband's	i	67
Beautiful	ii	29
Bengal	i	57
Blood-Rumped	ii	23
Blossom-Headed	i	63
Blue Bonnet	ii	21
Blue-Winged	ii	37
Bourke's	i	107
Elegant	i	83
Golden-Crowned	ii	16
Half-Moon	ii	88
Javan	i	87
King	i	93
Many-Coloured	ii	26
New Zealand	ii	13

INDEX.

	Vol.	Page.
Parrakeet, Orange-Bellied	i	103
Paradise	ii	29
Passerine	ii	37
Pennant's	i	141
Red-Rumped	ii	23
Red-Winged	i	71
Ring-Necked	i	57
Rose-Hill	ii	1
Rosella	ii	1
Splendid	i	99
Swift	ii	33
Undulated Grass	i	111
Yellow-Rumped	ii	8
Parrot, Carolina	ii	84
Dusky	ii	110
Festive Amazon	ii	94
Green Ground	i	121
Grey	ii	41
King	i	93
Red-Vented	ii	108
Senegal	ii	57
Violet	ii	110
Passerine Parrakeet	ii	37
Pennant's Parrakeet	i	141
Psittacus Alexandri	i	87
Amazonicus	ii	98
ararauna	ii	75
atricapillus	i	45
aurantius	i	103
aureus	ii	88
auriceps	ii	16
Barrabandi	i	67
Bourki	i	107
canus	i	135
Carolinensis	ii	84
cyanocephalus	i	63
cyanopygus	i	93
discolor	ii	33
Dufresnei	ii	96
elegans	i	83
erithacus	ii	41
erythropterus	i	71
eupatrius	i	51
eximius	ii	1
festivus	ii	94
flaveolus	ii	8
Goffini	i	1
hæmatogaster	ii	21
hæmatonotus	ii	23
hyacinthinus	ii	61
Leadbeateri	i	13

	Vol.	Page.
Psittacus leucolophus	i	7
leucotis	ii	92
Le Vaillanti	ii	104
macao	ii	69
maracana	ii	81
menstruus	ii	108
militaris	ii	65
multicolor	ii	26
nasica	i	27
Novæ-Hollandiæ	i	33
Novæ-Zealandiæ	ii	13
palliceps	ii	6
passerinus	ii	37
Pennanti	i	141
pezophorus	i	121
pulchellus	i	77
pulcherrimus	ii	29
pullarius	i	125
roseicapillus	i	21
roseicollis	i	129
Senegalus	ii	57
splendidus	i	99
sulphureus	i	17
Swainsonii	i	39
torquatus	i	57
undulatus	i	111
violaceus	ii	110
Purple-Capped Lory	i	45
Red and Blue Macaw	ii	69
Red-Rumped **Parrakeet**	ii	23
Red-Vented Parrot	ii	108
Red-Winged Parrakeet	i	71
Ring-Necked Parrakeet	i	57
Rose-Breasted Cockatoo	i	21
Rose-Hill Parrakeet	ii	1
Rosella, Mealy	ii	6
Pale-Headed	ii	6
Rosella Parrakeet	ii	1
Rosy-Faced Love-Bird	i	129
Senegal Parrot	ii	57
Slender-Billed Cockatoo	i	27
Splendid Parrakeet	i	99
Swift Lorikeet	ii	33
Swift Parrakeet	ii	33
Turquoisine	i	77
Undulated Grass **Parrakeet**	i	111
Violet Parrot	ii	110
West African Love-Bird	i	125
White-Eared Conure	ii	92
Yellow-Rumped Broadtail	ii	8
Yellow-Rumped Parrakeet	ii	8

END OF VOL. II.

B. FAWCETT, ENGRAVER AND PRINTER, DRIFFIELD.

www.ingramcontent.com/pod-product-compliance
Lightning Source LLC
Chambersburg PA
CBHW032151160426
43197CB00008B/867